Fermentation of Rice Beverage

Fermentation of Rice Beverage

A Case Study

Dr. Chamgongliu Panmei

AKANSHA PUBLISHING HOUSE
NEW DELHI • GUWAHATI • VISAKHAPATNAM

AKANSHA PUBLISHING HOUSE
4649-B/21, Ansari Road
Darya Ganj, New Delhi - 110 002
Ph.: 23263193 / 9811582579 / 09435116718
Fax: 011-23263193
Email: ektabooks@yahoo.com
www.akanshapublishinghouse.com

Fermentation of Rice Beverage: A Case Study

First Published 2011

ISBN 978-81-8370-257-7

PRINTED IN INDIA

Published by Akansha Publishing House, New Delhi and Printed at Harmain Offset Press, Delhi.

Dedicated

to

my dearest

Mother and Father.

Contents

Preface

Even though fermentation has been known in every human civilization there are great disparities in technological know-how of fermentation processes. Today, many of the indigenous traditional foods have been even termed as backward or poor peoples's food though there are many positives aspects to their credit. Fermented foods are part and parcel of Manipuri food habits, and in the cultural history of early Manipur, many communities used alcohols and rice-beers as part of their diet and in their socio-religious practices. In alcoholic fermentation, rice is used as a sole raw materials or source of alcohol and *Albizia* bark as ingredients. It is belief that used of *Albizia* bark in the fermentation enhanced the product as well as the quality of the product but the exact roled of the bark in fermentation were not known. This book which is the result of my Ph.D. work contains my research work on the traditional alcoholic fermentation of rice using *Albizia* bark. This book embodies four chapters. The methodology of tools and techniques with updated latest references has been given due place in these chapters. Figures and illustrations are presented to elucidate the topics making the subject more interesting and knowledge – rich. Topics like biochemical compound of the ingredients, microbiology; its application in fermentation, characterization of micro-organisms, standardization of the alcoholic fermentation process are covered in this book. This book depicts the role of ingredients particularly the *Albizia* bark in the alcoholic fermentation of rice.

I have published two papers out of this work in *Journal of Food, Agriculture & Environment (JFAE)*, 2003, Vol.1(3&4) and 2007, Vol.5(3&4). I would like to acknowledge the Editors of WLF Science and Technology, Helsinki, Finland for their kind publication of my papers in their journal.

In bringing out this book I would like to express my profound gratitude, and regard to my supervisor Prof. G.A. Shantibala Devi, Department of Life Sciences, Plant Pysiology, Manipur University, Canchipur, for her guidance and suggestions throughout the course of my work

My sincerest gratitude also goes to Dr. A .K Sharma, Head, Food and Technology Division, Bhabha Atomic Research Centre (BARC), Trombay, Mumbai for his meticulous guidance and also for providing Laboratory facilities in Food and Technology Division (FTD), BARC during the course of this research. I am also grateful to Scientific Officers; S. Gautam and Dr. Variyar S. Prasad, Senior Scientist, Dr. K.K. Surendranathan, FTD, for their assistance and guidance. I would like to acknowledge the generosity of the authority of the BARC, Government of India, Trombay, for allowing me to carry out the research work in their centre.

I appreciate IMTECH, Chandigarh for providing facilities for the microbial identification.

I thank my beloved parents, brothers, sisters, my husband Dr Ram Kamei, my father-in law Prof Gangmumei Kamei for their valuable support and encouragement in publication of this book. Finally my sincere thanks to Akansha Publishing House, New Delhi for publishing this book.

Most of all I thank the Almighty God for his love and blessings.

Imphal **Champongliu Panmei**

1
Introduction

Fermentation is one of the oldest methods of food preparation employed since the ancient times. It is the oldest form of biotechnology essentially consisting of transformation of the simple raw materials into amazing range of value added products by utilizing the phenomenon of the growth of microorganisms or their activities and various substrates. The word 'fermentation' is derived from a Latin verb **'fervere'**, which means to boil. The conventional definition of fermentation is the breakdown (metabolism) of larger molecules, like carbohydrates, into simple ones under the influence of microorganisms for their enzymes. In a microbiological way, fermentation is defined as "any process for the production of useful products through mass cultures of micro-organisms", whereas in a biochemical sense it means the numerous oxidation-reduction reactions in which organic compounds, used as source of carbon and energy act as acceptors or donors of hydrogen ions. The organic compounds used as substrate give rise to various products of fermentation, which accumulate in the growth medium (Riviera, 1977).

Fermentation is a gradual change brought about by the enzymes of some bacteria, moulds and yeasts in which the carbohydrate or like compound are broken down under aerobic or anaerobic conditions to produce energy in the form of ATP. Varieties of food and beverage such as bread, beer, cider, wine, cheese and yogurt are made via the fermentation process. Fermented food substances are food substrate that are invaded or overgrown by edible microorganisms whose enzymes, particularly amylases; proteases (and lipids) convert to nontoxic products with flavours, aroma and textures that are pleasant and attractive to the human consumer. It is such a special class of food where, during its production, a virtually complex mixture of protein, fats and carbohydrates undergo modification and the ultimate product is used directly as a food, or its additive or as its basic ingredient. Fermented food of plant or animal origin, prepared mainly with the intention of adding variety to an otherwise monotonous diet are a favourite dish of the Oriental kitchens. Textural, nutritional and sensory modifications could be achieved in the original food materials through the use of microorganisms, which grow either naturally or are inoculated during the preparation process of protein and calories. Some of the advantages of fermentation are that it helps to add flavour to a bland cereal-vegetable diet and prevent spoilage of food, prolongs shelflife (Campbell-Platt,1994) and improves the digestibility of ingredients for human consumption. It contains vitamins and serves as important source of proteins and calories.

Human beings have been using this technique from the time of very early civilisations and the recorded evidence is in the Babylonian brick from 2800 B.C. with a recipe for fermented barley. The processed fish sauce of today has found its roots in the Roman Empire, including both Europe and North Africa. The classified fermented foods are beverages, cereal products, diary products, fish products and miscellaneous products. Diary products constitute the largest numbers of fermented foods, while the most lucrative is in beverages sector comprising wines and liquors. Fermented cereals and pulses along with millets have been the staple food in Asian and African subcontinents from the early times.

The origin of brewing reaches far back into the prehistory of man. Artifacts taken from the mines of ancient cities such as Mesopotamia indicate that the formal practice of brewing existed well over 5,000 years ago. Writings and drawings recovered from ancient Egypt give details of commercial brewing and distribution of beer. It was used as a mortuary offering in the 5th Dynasty about 2800 B.C and later it appears to have been the national beverage. It also played an important part in religious worship and some of the festival offerings. Ancient Aryans in India produced a beer-like beverage called 'Sura', made from grain as early as 2000 B.C. and in China a kind of beer called 'Kiu' was produced. Monasteries and other establishment brewed beer during medieval times and produced as much as we know it today using malted barley, water hops and yeast. Greeks learned about brewing from the Egyptians, Romans learned about beer from Greeks and then Gaul and Spain. The Northern European races probably discovered the techniques of brewing long before the Christian era. The earliest Teutonic and Celtic beverages were made from a mixtures of corn and honey and hence approximately mead. There is no direct evidence of beer brewing in Britain prior to the Roman occupation but by the Anglo-Saxon times, probably as the result of invasion, beer or ale was well established. The Indians, long before the advent of Europeans, made various alcoholic beverages; but the history of American brewing really begins in 1584 when the British brewed beer from corn (maize) during their first attempt to colonize Virginia. Later in 1612 Dutch colonists are said to have set up a brew house in the most southerly point of Manhattan Island.

Amongst the fermented products, wine has been consumed throughout the ages as food and food adjunct and there is no doubt that *wine is a healthful beverage.* Though the virtues of wine have been exploited in folklores and in medical arts and sciences, the experimental evidence in its favour came only recently.

People of Asia are the pioneers in the developmental mean of fermenting vegetable proteins to produce meat-like flavour. Indonesians developed the fermentation methods for introducing meat-like texture in vegetable products. Koreans are credited for production of acid-fermented vegetable. People of Egypt developed wheat breads leavened with yeasts while Indians discovered methods for souring and leavening cereal legume batters.

Man has exploited a huge number of microorganisms in various fermentation processes for production of a wide spectrum of products including food fuel, chemical and fuel and other products. Over the past several decades, fermentation industry has developed dramatically and now it has reached a new dimension with the era of microbial biotechnology. The biological agents involved in food fermentations include filamentous fungi, bacteria and yeasts occurring in succession or in combination. In general, filamentous fungi bring about the saccharification and proteolysis of starting materials; bacteria cause acidification and leavening while yeasts ferment the sugars to produce alcohol and esters which impart desirable flavour to the products.

No exaggeration, alcoholic fermentation was the first fermentation known to the mankind. It is the anaerobic biological degradation of glucose and fructose to ethanol and carbondioxide with a number of side products. From ancient times people all over the world have developed traditional processes for production of alcoholic beverage (wine) either from natural sugars such as in wine or by degrading the starchy materials (beer). Fruit wines referred to as 'soma' (in

Hindu scriptures) have been part of food of man and were used as medicine too. Distillation was however, a later development. Wines are made principally from fruits like grape or extract of other fruit of various types, like cherries and berries. The characteristics and distinctive flavours of wine are partially determined by the variety of fruit used, environmental conditions of the grape plant and the parts of the fruit used while the type of organisms used influence the quality of the wine. Traditionally the yeasts used to ferment the juice were those occurring naturally on the surface of the fruit. Grape juice (27 % sugar) is fermented into alcohol by various strains of Saccharomyces ellipsoidues and also rendered the chemical constituents, which alter the flavours. A low acid content favours production of acetaldehyde, glycerol and volatile and fixed acids which have a detrimental effect on the flavor, color and stability of the wine.

The alcoholic beverages where cereals are used include a variety of beer-like beverage and distilled liquors called whiskies. These beverages are produced and consumed in large quantities throughout the world. Beer is produced after the fermentation of mixture of barley malt and starchy solution by saccharomyces cerevisiae (top fermenter yeast) or S.carlsbergensis (bottom fermenter yeast). Fermentation converts sugar into alcohol and also brings about minor chemical changes. Whisky is made from fermented grains (corn, wheat, barley, malt or rye-malt) cooked and saccharified with peated malt and fermented by top yeast. After fermentation, the culture fluid contains alcohol, traces of acids and ester. The broth is distilled in order to concentrate the alcohol to about 51 to 59 %. Rum is prepared from black strap molasses containing 12 to 14% fermentable sugar and fermented by top yeast (*S. cercvisiac*). Thc culture fluid is distilled to remove the alcohol and used as rum.

Ethyl alcohol, commonly known as alcohol is one of the most important and popular industrial fermented products.

Different cross-section of people from the chemical manufacturer to the housewife depends on alcohol in one way or the other. There was a sudden boom in the use of industrial alcohol during World War II due to its application in synthesis of rubber and smokeless powder. The ever-reducing level of fossil fuels and the continuous rise in its market price have led the scientists to find out a viable long-term substitute for liquid petroleum.

Ethanol is a liquid fuel alternative to automotive fuel and non-petroleum fuel option which can be produced from renewable resources. The production of ethanol by fermentation is a proven technology. Moreover the technology of using alcohol with gasoline in spark-ignition engines has also been developed very recently. In fact, a mixture of 90 % gasoline and 10 % ethanol is being marketed in some countries. These reveal that there is a growing interest in ethanol as a viable liquid fuel. Any raw material containing hexose sugar or materials that can be converted into hexose sugar can be used as raw materials for fermentative ethanol production. Cereal grains, starch root, plants and some cacti are the starting raw materials for ethanol fermentation. The fermentation of starch from cereal grains has been practiced for a long time. Biotechnologists have focused the attention on the production of ethanol from various sources (substrate) like corn, potato, rice barley cassava roots, wheat, cereals and other grains, saccharine materials, cheese waste, coffee waste, potato peel waste, pineapple waste, orange peel waste and apple pomace.

The Indian population depends heavily on several fermented products, many of which are also made from combinations of cereal and legumes grains. This gives higher protein content and provides a better-balanced ratio of amino acids. Indian fermented foods like dosa and idli are presently available in all metros across the country, while specially foods specific to regions as dhokla appam, millet porridge, khaman, kenima, papadam, jelebies etc. are also popular. Indian ruhi is a strong alcoholic beverage made by fermentation of boiled rice. It is a social drink produced primarily by tribal people in the Naga inhabited areas and other hilly regions of eastern India. Indian madhu is a primitive low-alcohol rice wine in which raw rice soaked in water is used as the substrate. This is also a tribal drink produced and consumed primarily in Naga inhabited areas and in the hills of North Eastern India. Indian rice beer pachwai is also a popular drink manufactured from steamed rice by adding powdered 'bhakhar'. Chowan is a popular fermented product of Tripura used for preparing beer; Appup in Arunachal Pradesh is also very popular. A variety of kiad (liquor) is prepared from rice or millet in Meghalaya. Iad thang is obtained by means of heating (distillation). Iad pyrnah is another kind of kiad specially prepared liquor used for religious ceremonies at 'Sutnga'. Iad laroh- a kind of Jaintia liquor is used for entertainment and ceremonies like erection of stone monuments. Beer brewed from rice or millet is consumed on all occasions in Garo villages. According to the customs, belief and common saying of the Garos, a child who drinks no rice beer will not have blood!

Rice beer is a common drink consumed in large quantities on festival and feast by the tribals in the North Eastern part of India. In Mizo society drinking of zu - a Mizo beer was very common and it was part and parcel of all ceremonies and not a single ceremony is performed without zu. Kukis use zu in customary marriage proceedings, rituals and ceremonies. Zu, the most important traditional drink, occupies an almost indispensable place in the social and cultural life of Zomis. Rice beer is a common drink on the festivals and feast for the people of Kachcha Naga (Liangmei), Zeme and Rongmei (Kabui) tribes; they also consistently drink certain quantity of rice beer daily, and thus making it an integral part of their food habit. A primitive, peculiar and harmless drink called 'zou-ngau' consisting of pounded rice mixed with boiling water, brought into fermentation by addition of malted paddy is very common among the Zeme, Liangmei and Rongmei tribes. **'Zu'** or rice beers are variously flavoured, according to their experience and preferences. One common variety is very sour, pleasant enough but thin. The Tangkhuls make a thicker variety which is yellowish in colour and sweeter in taste; they use wild herbs for flavouring. Kukis often use the leaves of the datura plant and stramonium, which account for the strength of their brew.

In the cultural history of early Manipur it is found that wine industry was widely prevalent as people (of both sexes) were very fond of drinks. Wine played a great role on the socio-economic and religious activities of the people. Even today few tribals and the Loi villages of Sekmai, Andro, Phayeng and Leimaren manufacture liquor on a large scale. A group of people called 'Lois' of Manipur (Andro, Khurkhul, Phayeng and Sekmai) is well known for their preparation of alcoholic drink called **'yu'** or **'kalei'** which is a popular distilled product. As a traditional custom among the Lois, liquor is an essential item for 'Yupanthaba' or 'Ipanthaba'. **Yu** is used and drank in common festivals, ceremonial rites, marriages, and religious rituals at birth and death. From the very beginning, Andro is popular for its alcoholic drinks and distillery work. About 90% of the families are engaged in this traditional profession. This work is generally operated by female section of the society. In a festival called 'Haochongba' the potent spirit distilled at Sekmai or Andro is served with a sumptous feast. The Meiteis, after their adoption to Hindus have abandoned and discarded the drinking of country liquor (particularly the rice beer). Christian converts convinced by their experiences and moral convictions have also abandoned this practice. **'yu'** or **kalei** is used by practioners of traditional medicines (Maiba and Maibi) in the preparation of herbal concoction. Such concoction is used to treat different ailments. It is a common saying that a person who regularly consumes *'Zu'* (rice beer) has a high body resistance to diseases. One of the unique features in the preparation of *'yu'* (country liquor in Manipuri) is the use of *Albizia* bark (*yangli* in local language) powder in the process.

Albizia is a leguminous plant belonging to mimosaceae family. They are usually trees and rarely shrubs or climbers. About 30 to 50 species are spread throughout the tropics of the old world. It is distributed in North Australia, South West Africa, Tropic Africa, Abyssinia, Eastern and Central Asia, China, Japan, Ceylon, Nepal, Burma, Philippines, Malay isles. In India, it is mainly distributed in Central and Eastern Himalayas, Bihar, Western Peninsula, Khasi and Jaintia Hills, Garo Hills, Assam, Manipur and Nagaland. Albizia spp. is a native plant of Manipur grown scatteredly in the wet and semi-evergreen forest of the hills. *Albizia lebbeck, A.myriophylla, A. kalkora* and *A. samman* are the few other species found in the same area. *A. julibrissin, A. lebbeck, A. anthelmintica, A. fastigiata, A. gummifera, A. antunesiana* and *A. lophantha* are some of the species used as medicine in countries like China, Indochina, Annam, Abyssiana, Madagaskar, East Africa, Southern Africa, Southwest Africa and Northern Rhodesia etc. *A.myriophylla* Durazz is used as a substitute for 'licorice' in Vietnam. *A.myriophylla* Dur. and *A. kalkora* Dur. are used as ingredient in the preparation of **'yu'** in Manipur.

In the preparation of *'yu'* or ***'kalei'***, a starter culture called hamei is first prepared from the flour of raw rice mixed with dry bark powder ***yangli*** (*Albizia* plant). The mixture is moistened with water and inoculated with dry ***hamei*** powder from the previous batch. The cake which may be 3 to 5 cms in diameter and 1 to 1.5 cms thick when flattened are placed on a bamboo tray and incubated for 4 to 5 days in ambient temperature, followed by dehydration to preserve the cakes until needed. ***Hamei*** cake preserves the essential microorganisms for several months at room temperature. Secondly, rice is cooked and spread on a mat and the crushed ***hamei*** cake is sprinkled over the spread rice and thoroughly mixed. The mash is then place in a large earthen pot covered with cloth and incubated for 5 to 6 days until the mixture froth and slight liquefaction of the rice starch occurs. More water is added and the mash is transferred into a large aluminum vessel, which has a rounded hole on the side and heated on fire. Cold-water pot is placed on top to obstruct the rising vapour. The condensed vapour is collected through the pipe from the side hole of the pot and thus **'yu'** is finally prepared.

People of Manipur are still confined to this traditional **'yu'** manufacturing procedure and they are in the belief of existence of goddess of **'yu'** for high and quality yeild. The work report here is designed to test a standardised laboratory scale procedure for **'yu'** production and to study the effect of the raw materials used on product quality and quantity. Therefore, the topic "Studies on the role of *Albizia spp.* in the fermentation process of different rice cultivars" has been chosen for the present research work.

2

Review of Literature

The process of fermentation became known to human being in the early civilizations. The phenomenon was viewed scientifically only during 1850's by Louis Pasteur. Later, Buchner and Buchner (1897) revealed that cell-free extract of yeast could replace living organisms to carry out the fermentation. For centuries, diverse plant and animal materials have been fermented by various bacteria, yeasts and fungi (Hesseltine and Wang, 1967). However the kind of microorganisms used in traditional fermentation are restricted to a relatively few genera, including *Aspergillus, Rhizopus, Mucor, Actinomucor, Monascus, Saccharomyces, Neurospora, Acetobactor, Bacillus* and *Lactobacillus.*

Preparation of many other indigenous products is still a household art where fermentation is induced by uncontrolled inoculation, which produces variations in quality and stability of the product (Sanni, 1993). Indigenous cereal-based fermented products have aroused interest among nutritionists and food scientists and several studies have been reported on the monitoring of physical, biochemical and nutritional changes during fermentation (Chavan and Kadam, 1989; Steinkraus, 1997; Ottogalli and Galli, 1997). Bread and beer are the most popular cereal-based fermented foods, but there are also many other indigenous cereal-based fermented foods prepared in different parts of the world, such as **tempeh** (Steinkraus, 1977; Ayres et al., 1980; Beuchat, 1995), **pito** (Steinkraus, 1997), **sekete** (Sanni, 1993; Adegoke et al., 1995), **kwass** (Pederson, 1979) and **boza.**

Reviews on fermented foods and drinks using rice products include those by Hesseltine (1965); Wang and Hesseltine (1970); Wood (1981); Steinkraus (1981,1983); Chiao (1981); Batra (1981); Lizuka and Lin (1981); Nunokawa (1981); Kodama (1986); Konishi et al. (1989) and Wang (1980,1989). Van Veen and Steinkraus (1970) have briefly summarised the nutritive value of fermented foods derived from rice. Fermented rice-food products can be classified into three categories (Wang, 1980): solid, paste and liquid. The solid state fermented products include starter types such as **p'eka** (pekka), **anka** (china), **ragi** (Indonesia), **koji** (Japan), predigested "Yellow rice" (Ecuador) (Van Veen and Steinkraus, 1970), and breadlike foods (Steinkraus, 1983). Characteristic paste products are **miso** (Japan) and **chiang** (China) for seasoning foods, some sweet/sour alcoholic foods (Steinkraus, 1983) and **tieng-chiou-niang** (Wang, 1980). **Shao-hsing** wine (China), **sake** (Japan) and rice vinegar are well-known fermented rice foods typical of the liquid form (Hesseltine, 1965; Wang, 1980). It appeared that fermented grains retained their consistency much better than unfermented ones (Herzfeld, 1957).

In the Orient rice is widely used to make wine and vinegar, although in the West, it is used only as an adjunct to the production of beer. Modification and improvement of the processes have been described by Liu et al. (1959 a-c; 1960); Wang et al. (1970), Lin et al. (1972), Murakami (1972), Hayashida et al. (1972), Yoshizawa (1977), and Kodama (1986).

Nago et al. (1998) studied the characterization, physiocochemical and microbiological aspects of the Beninese traditional **ogi**, a fermented maize sherry.

Gotcheva et al. (2001) studied on monitoring the fermentation of the traditional Bulgarian beverage **boza.**

Pederson (1979) also reported **busa** as an acidic alcoholic drink prepared from millet, rice and sugar and consumed by the Tatars of Krim, Turkestan and Egypt.

Bonza is a fermented alcoholic beverage consumed in Egypt and prepared from wheat or maize (Baervald, 1988; Sanni, 1993). It is a fermented alcoholic wheat beverage that has been known since the time of Pharaohs (Marcos et al., 1973;Marcos, 1977)

Kenyan **busaa** is described as an acidic alcoholic beverage with a smooth texture made from maize, finger millet and 'sorghum' (Steinkraus, 1977; Baervald, 1988; Sanni, 1993).

Chicha is a clear yellowish, effervescent alcoholic maize beverage with a flavor resembling cider, and has been consumed by the Andes Indians for centuries. Its alcoholic content varies from 2 to 12% v/v. (Escobar, 1977; Escobar et al., 1977). In the days of the Incas, **chicha** was considered to be " the vehicle that linked man to his gods through the fecundity of the earth" (Nicholson, 1960). **Chicha** was used to induce the "thunder god" to send rain and was also used in 'sun' and 'harvest' festivals (Maxwell, 1956). Variety of procedures exists for the production of **chichi** (Nicholson, 1996; Cutter and Cardenas, 1947) and the methods of extracting the salwated or germinated maize flour vary rather widely. Gomez (1949) found that various yeasts like *S.cerevisiae, S.apiculata, S. pastorianus* and bacteria of genera *Leuconostoc, Lactobacillus Acetobacter* and various moulds including *Aspergillus* and *penicillium* were present. Escobar (1977) studied the **chicha** fermentation.

In Mozambique women chew the **yucca** plant, spit it out and allow it to ferment to a beer called **masata** (Pederson, 1979).

A number of studies have been made of sorghum (*Kaffir*) beer production of South Africa (Aucamp et al., (1961); Horn and Schwartz (1961); Von Holdt and Brand (1960 a, b); O' Donovan and Novellie (1966; 1959, 1960 a, b; 1962 a, b, c; 1966 a, b, c, 1968,1976). The essential steps in brewing are malting, mashing, souring, boiling, conversion, straining and alcoholic fermentation (Schwartz, 1956). Platt and Webb (1946) concluded that the major Saccharification was not due to malt amylases but due to moulds growing on the grain. Aucamp et al. (1961) reported chemical analysis of municipal 'Kaffir' beers. Vander Walt (1956) described the microbiology of ' Kaffir' beer fermentation.

Starch is a very important component in **ban tu** beer which contains a considerable amount of both gelatinized and ungelatinized starch (Novellie, 1966 b; Novellie and Schuffe, 1961). Most of the amylolytic breakdown occurs when the mash is in the pH range of 5 to 4.0 (Novellie, 1966 b).

Tesquino -a slurrylike alcoholic beverage prepared by fermentation of germinated maize or maize-stalk juice is studied by Taboada et al. (1977). The yeast *S. cerevisiae* is an important microorganism involved in the alcoholic fermentation of **tesquino** (Herrera and Ulloa, 1973).*Bacillus megaterium* has been isolated from **agave tesquino** (Ulloa et al., 1974).

Nigerian **pito** is a light brown, sweet-sour and slightly bitter alcoholic beverage with a fruity flavor made by fermentation of malted mashed maize or sorghum (Ekundayo, 1977). During malting and subsequent processing, amylases from the germinated grain hydrolyse the starches to fermentable sugars, principally maltose and glucose. The moulds on the grains which may also contribute amylases during germination have been studied by Novellie (1960 a, b) and Faparusi (1970).

Talla (tella) is an Ethiopian home-processed beer with a smoky flavor and a tan to dark brown colour (Vogel and Gobezie, 1977). It is also frequently used as a medicine (Tegegne, 1957) and serves as a solvent for tapeworm medicine and other drugs.

Mbege is made from malted millet and acid-fermented banana juice. It is consumed daily, primarily by people living near Mount Kélimanjaro (Harkishor, 1977).

Zambian **opaque** maize beer, the usual alcoholic beverage of farmers and low-income urban workers, is a brown colored, viscous and slightly bitter beverage made by adding cereal malt to a maize porridge. Lovelace (1977) reported that **monkoyo** root extract contains some sugar-reducing agent and an active amylase and **munkoyo** (maize slurry) contains an average of 0.43% ethanol w/v.

Sake is a clear, pale yellow rice wine with an alcoholic content of 15 to 16% or higher with a characteristic aroma, little acid and slight sweetness (Murakami, 1972). In very early times, the rice for **sake** was saccharified by chewing it (Kodama and Yoshizawa, 1977). Raw material (water) used for **sake**, should be colorless, tasteless and odorless, neutral or weakly alkaline and substantially free of iron, (less than 0.02 ppm), nitrate, ammonia, organic substances, and harmful microorganisms (Kodama and Yoshizawa, 1977). Rice of the short-grained varieties is considered best for **sake** manufacture and large grain are considered desirable. Biochemical changes occurring in **koji** are á-amylase (liquifying amylase) and amyloglucosidase (Saccharifying amylase) which develop and hydrolyse the rice starch into dextrins, glucose and maltose. Acid and alkaline proteases develop and hydrolse the proteins to peptides and amino acids and B vitamins are synthesised. Initial starch content of rice is 90.4% and falls to 69.9%. Reducing sugars increase from 0.2% to 21.4% in **Koji** (Kodama and Yoshizawa, 1977). Under traditional fermentation conditions, *Pseudomonas, Achromobacter, Flavobacterium* or *Micrococcus spp* produce nitrites and nitric acid from nitrates in the water (Murakami, 1972). These are followed or possibly accompanied by *Leuconostic mesenteroides* **Var-sake** and *Lactobacillus* **sake** (Kodoma and Yoshizawa, 1977). *Hansenula anomala* yeasts are resistant to nitrate but cannot survive the anaerobic conditions in the mash. **Sake** yeast eventually propagates to levels of 3 to $4x10^8$ cells/g, at which time it makes up 95 to 98% of the microbial flora. Nunokawa (1981), Steinkraus (1983) and Kodama (1986) who extensively reviewed the treatments of water and rice grain on brewing of **sake** and Murakami (1972) described the characteristics of microorganisms and found **sake** yeast somewhat different from *saccharomyces cerevisiae*, while **koji** moulds were found to be definitely different from *Aspergillus flavus*. An average **sake** contains 4.2% w/v total sugar (as glucose), 3.46% w/v fermentable sugar (as glucose); 1.52 meg/100ml acid; total organic acids (mg/100 ml), 115.22; glutamic acid (mg/100 ml), 20.23; total nitrogen (% w/v) 0.0726; formal nitrogen (% w/v), 0.0288; and ethanol (% v/v), 15.0. **Sake** contains various higher alcohols and esters, n-propanol (120mg/ester); isobutanol (64mg/liter); isoamyl alcohol (170mg/liter); and 2 phenethanol (75mg/liter).

Honey wine has been an indigenous fermented beverage for thousands of years (Morse and Steinkraus, 1975). Brothwell and Brothwell (1969) suggest that half-empty honey pot left in the rain may have been the first alcoholic drink. Rate of fermentation of all honey can be increased by addition of nitrogen and growth factors for yeast (Steinkraus and Morse, 1966). Sorin (1921) analysed 42 honey wines produced in Eastern Europe and the alcohol contents ranged from 6.4 to 16.6% v/v with an average ethanol content of

11.0% v/v. Desta (1977) reported ethanol contents in traditional **tej** ranging from 13.18 to 13.73 v/v.

Sugarcane wine is known in Japan as **shoto sake** (Kozaki, 1976), in Philippine as **Philippine basi** (Sanchez, 1977; Tanimura et al., 1978b) and in Kenya as **Kenyan muratina** (Harkishor, 1977). Kozaki (1976) reports that the dominant organisms in **basi** are *Saccharomyces endomycopsic* and Lactic acid bacteria. Alcohol content is from 9.9 to 13.4% v/v.

The palm wine fermentation is always a lactic alcoholic-acetic one, involving lactic acid bacteria, yeasts and acetic acid bacteria as well as *Zymomonas* and *Micrococcus spp* (OKafor, 1972 a, b). The most desirable quality in palm wine is its sweet alcoholic flavor, but the slightly sharp taste produced by a combination of small amounts of lactic acid and acetic acid is also desirable (Okafor, 1977 b). The palm sap contents 14 to 18% sugar, 0.23% protein, 0.02% fat mineral matter, and ascorbic acid (5.7 mg/100ml) [Theivendirarajah et al., 1977 b]. Alcohol content ranges from 4.8 in pasteurised bottled-coconut **toddy** to 10.3 in fresh toddy (Merican, 1977) and the yields can be increased from 6.5 to 10%(v/v) using selected strains inoculated into prefermented sap (Samarajeewa et al., 1976 a).

Pulque is the national drink of Mexico where it was inherited from the Aztecs (Goncalves de Lima, 1975). The substrate for **pulque** is Agave juice. Ethanol producing yeast *S. Cerevisiae* and others viz: *Endomycopsis, Pichia,* and *Torulopsis spp.*, are found (Sanchez Marroquin and Hcpe, 1953). *Zymomonas mobilis sub sp. mobilis* is considered by some (Swings and Deley, 1977) to be the causal organism for the **pulque** fermentation. Herrera and Ulloa (1975) reported the presence of *Kloeckera apiculata* in fermenting **pulque.**

Mexican **mescal** is a tan-colored, sour beverage with an alcoholic content similar to beer. The substrate is the core of the Agave plant, which is stripped of its leaves, cooked and filtered to yield a sugary pulp. The pulp is inoculated with a starter from a previous fermentation (Pennington, 1969).

Jackfruit wine, an alcoholic beverage made by fermentation of jackfruit pulp fermented by yeasts resembles *Endomycopsis* (Dahiya and Prabhu, 1977). Alcohol content in the wine attains 7 to 8% v/v within a fortnight.

Urwaga is an effervescent slightly sour alcoholic beverage made from banana, sorghun, millet or maize fermented by lactic acid bacteria and yeasts (Harkishor, 1977).

Steinkraus (1983) summarised the literature on starter-making extensively at the 5th International Conference on Global Impacts of Applied Microbiology in Bangkok, Thailand, in November 1977.

Microbial starter cultures are used for inoculation of some fermented rice products. Wang (1980) reviewed the ancient histories (Methods of farmers, 1968) of Chinese starters, which are somewhat similar to those found in South-East Asia. Steinkraus (1983) described the starter cultures and the organisms found in **chinesische hefe** and **javanische ragi,** along with the amylolytic yeasts and molds found in the fermentation of **arak**. Other scientists who have studied and discussed these starter cultures include Hesseltine (1965), Ko (1981), Sie (1962) and Soedarsono (1972). Ellis et al. (1976) described the genus *Amylomyces* (formerly *Chlamydomucor),* which is extremely important in the production of fermented rice pastes and wines. Making the starter (**chu or ch'u**) involves a solid-state fermentation occuring in an open system in which moisture is lost by evaporation (Wang and Chiou, 1977; Wang, 1989). The size and texture of the starter determine the distance the gases move, the volume of gas and the water-holding capacity; it also affect the thermodiffusibility in the starter (Wang et al.,

1968; Lai, 1984; Lai et al., 1989) and the germination or growth of the native-occurring microbes. The effect of high partial pressure of respiratory carbondioxide inside the starter has been demonstrated in the nuclear number of the sporangiospores of Rhizopus in the solid-state mash of sorghum fermentation (Wang et al., 1970; 1974; Wang, 1978 a, b).

Recently, starter making in Taiwan has been improved with the use of a disk type of bioreactor for solid-state fermentation in place of the traditional wooden boxes (Wang, 1980). Strong aeration is provided from the bottom plate through fine holes and mixed with a special device, as described by Mudgeff (1986) and Toyama (1976).

Ragi murcha, loogpang, bubod, chin-chu, and Chinese yeast ball are used as starters from a number of fermentations based on rice and cassava in the Orient. The starter consists commonly of certain species of *Mucor, Rhizopus*, and *Amylomyces;* other moulds are not found even though the production of starter is often made under unsanitary conditions. All the isolates from these starters grow under anaerobic conditions (Hesseltine et al., 1985).

The cultures that produced favorable aroma showed anaerobic growth of *Mucor spp, Rizopus spp. Anylomyces spp.* (Wang, 1989; Wand and Kim, 1989).

The starter for **lao-chao** is **chiu-yuch** or **peh-yuch** (Wang and Hesseltine, 1970). Major species present are *Rhizopus oryzae, R. chenénsis* and *A. rouxii.*

Some of the paste products also have a liquid component that can be collected and consumed as beverage, as described by Steinkraus (1983). **Tieng-chiou-niang** is a related product (Wang and Hesseltine, 1970; Wang, 1980).

Tape ketan and **tape katella** is made of glutinous rice and smashed cassava (Ko, 1965; 1972). A detailed account of the preparation of **tane** (seed) **koji** is found in the review of miso fermentation (Shibasaki and Hesseltine, 1962).

Recently, Chon and Hwang (1989) and Hwang (1988) studied micro flora and enzyme activities of **tou-pan chiang**.

Nakano et al. (1976) have described the mixing ratio of rice, soybean, and salt in making **miso shibasaki,** and Hesseltine (1962) found that besides Aspergillus, other bacteria like lactic, osmophillic and halophillic bacteria, and yeasts are also involved in **miso**.

The starter for Korean **yakju** and **takju** is **naruk** (Park et al., 1977). In two samples, Kim (1968) isolated per gram of sample 2.0x10 to 8.4×10^7 colony forming units (CFU) of *Aspergillus oryzae*, 1.6x10 to 2.9×10^7 cfu of *A. niger*, 2.0×10^6 cfu of genus Rhizopus, 1.3×10^7 to 2.6×10^7 bacteria, 5.0×10^6 to 2.0×10^7 aerobic bacteria and 1.4×10^5 to 6.0×10^5 yeasts.

Wehmer (1900 a, b) described the starter cultures and the organisms found in **chinesische hefe** and **javanische ragi** went while Prinsen Geerligs (1895) investigated the amylolytic yeasts and moulds found in the fermentation of **arak**. Other scientists who have described and discussed these starter cultures include Heyne (1950), Hesseltine (1965), Ko (1965; 1972), Sie (1962) and Soedarsono (1972). Ellis et al., (1970) described and discussed the genus *Amylomyces* which is extremely important in the production of alcoholic rice pastes and wines.

Ragi and Chinese yeasts are not food in themselves but are starters consisting of rice powder with certain species in which the proper organisms are present. In China and South East Asia, this powder is used to start food fermentations. Two food items made from this starter are the Indonesian **tape** (Djien, 1972) and the Chinese dessert **lao-chao** (Wang et al., 1970). A recent study of **ragi** by Hadisepoetro et al. (1979) reported counts of bacteria up to 5.0×10^4 and identified the 17 strains isolated from **ragi** as *Pediococcus pentosaceus*.

The spices added to some **ragis** contribute other microorganisms or may inhibit development of considerable microorganisms (Sie, 1962; Saedarsono, 1972). Raw dry spices contain a wide variety of microorganisms. Essential oils found in species have some inhibitory properties and although most species are not considered to be very bacteriostatic, they may, in combination with other materials in the foods exhibit a specific effect.

Under the condition given for production of **ragi**, amylolytic yeasts and moulds present on the species may be favoured, while growth of indesirable bacteria is inhibited.

Besides the microorganisms and starch rice powder, certain spices consisting of garlic, ginger, langkaus, *Citrus aurantium L.*, and pepper are always incorporated (Soedarsono, 1972). Soedarsono (1972) reported that garlic greatly inhibited growth of *Aspergillus niger* and *Bacillus subtilis*. On the other hand, *Rhizopus* and *Amylomyces* were only slightly inhibited by garlic and therefore, would grow in the **ragi** cakes. Ginger likewise did not inhibit *Amylomyces* or *Mucor*. Garlic, cinnamon and cloves are known to have antimicrobial activity (Frazier, 1958). Kissenger and Zaika (1978) reported that black pepper, all spices and nutmeg stimulated the production of acid by *Lactobacillus plantarium* but *Pediococcus cerevisiae* mixtures but did not stimulate bacterial growth.

Natural cereal-based fermentations are usually induced by yeast, lactic acid bacteria and fungi, forming sometimes, a complex population acting in concert. Yeast mainly degrades the carbohydrates while bacteria show a proteolytic activity (Chavan and Kadam, 1989). Fermentation by yeast and lactobacilli also affects the fat, mineral and vitamin contents of the media (Khetarpaul and Chauhan, 1991). Any species of Lactobacillus and Pedioccocus have been claimed to be associated with brewing (Emies, 1969; Eschenbecher, 1969)

Food of the Indian **idli** type, acidified and leavened through fermentation by heterofermentative lactic acid bacteria, constitute a very interesting group of cereal-based food of considerable potential and importance in the developing and also the developed world (Steinkraus, 1983; Batra, 1981). Leavening of **idli** is produced by bacteria rather than by yeast activity (Hesseltine, 1965; Steinkraus, 1983).

Recently, Chon and Hwang (1989) and Hwang (1988) studied micro flora and enzyme activities of **tou-pan Chiang**.

A total of 46 different strains of Rhizopus species were isolated from Indonesian **tempe** samples, soaking water and from commercially distributed rice meal **tempe** inocula (Baumann, 1992; Hering, 1991).

The dominant microorganisms of **ogi** were lactic acid bacteria and yeasts. The association of these two categories of organisms has been noticed in the spontaneous fermentation of **ogi** and many other cereal foods (Akinrele, 1970; Nout, 1980; Adegoke and Babatola, 1988; Hounhouigan et al., 1993). Steeping duration and temperature can induce dramatic changes in the microbial population of **ogi** (Akinrele, 1970). 54 strains of yeast were isolated from Beninese **ogi.** *Candida* species represented 41% of the isolates, including mainly *C. humicola* and *C. Krusei. Geotrichum spp.* accounted for 26% of the yeast isolates. Other isolates were identified as *Cryptococcus* and *trichosporon* species. Some of these yeasts were detected in Nigerian **ogi** (e.g. *C. Krusei, Rhodotorula Spp.*) by Akinrele (1970) and in **mawi** (e.g. *C.Krusei, C. Kctyr*) by Hounhouigan et al. (1993). Other yeasts like *Saccharomyces cerevisiae* and *Candida mycoderma* were reported.

Soybean yogurt is an example in which two strains of *Lactobacillus acidophilus* are employed with each strain contributing to the final product (Kanda et al., 1976). Ellis et al. (1976) found *Amylomyces* and its single species *A. rouxii* to occur in starter culture.

All the fungi used in oriental fermentations are those that grow first in the succession of moulds growing on plant material in nature. They rapidly use simple sugars. Before preservatives were used in bread, *Rhizopus stolonifer* was known as the "black bread mould" and *Neurospora* as the "peak bread mould". They were the first to grow in the microbial succession. Fungi in the order of *Mucorales* appear to be the ones domesticated in the tropics, whereas, *Aspergillus* seems to be used more in temperate or semitropical countries such as China and Japan. The relative humidity of the areas may be an important selective factor. The yeasts are represented mainly by the genera *Saccharomyces, Canadida (Torulopsis),* and *Saccharomycopsis* (Hesseltine, 1979).

Fermented foods involve relatively few Mucor species in the *Sphaerosporus* and *Racemosus* sections. Zycha et al. (1969) recognise *M.jacemosus, M.jraini, M.javanicus, M.jouxii* and *M.circinelloides.*

Actinomucor - the monotypic genus is known to be used in the **sufu** fermentation in which **tofu** is surface-moulded with *A.elegans* (Eidam) (Banjamin and Hisseltine, 1957).

Species of *Aspergillus* are commonly used in the orient to produce **koji. Koji** functions primarily as a source of enzymes to act upon starch, fats and protein in the fermenting substrate and it has a high tolerance to high temperature (Rapper and Fennel, 1965).

Ho (1976) examined 71 cultures of *Neurospora intermedia* – the fungus used in **ontjom**. A recent extensive study by Perkins et al. (1976) of *Neurospora* strains (675) collected all over the world shows *N.intermedia* to be abundant in Asia.

Batra and Millner (1974) study the yeast occurring in fermented food in India.

Djien (1972) isolated *Endomycopsis burtonii Boidin* from **ragi** and from rice ferment by Lodder (1970). Batra and

Millner (1974) report isolation of *Hensenula anomala* from **murcha** (similar to **ragi** but produced in Nepal and India) and in **kanji** (a beer-like beverage made in India from carrots and beets with species).

Sugar fermentation was ascribed to vital activity of yeast (Cagniard latour, 1836) and the fungal nature of yeast (Schwanm, 1837).

The essential microorganism for production of Chinese red rice (**anka**) is *Mucor purpurcus* went (Hesseltine, 1965).

Saccharomyces cerevisiae, S.chivalieri, Candida edax, C. Valida, C. tropicalis, C.krusei, Pichia membranaefaciens, Geotrichum candidum, Lactobacillus plantarum, Leuconostoc mesenteroides and *Micrococcus spp* have been frequently isolated from Ghanian wine (Nyako, 1977). *Zymomonas mobilis* has also been isolated from African wines (Swings and Deley, 1977).

Only one species of yeast, *A. saccharomyces sp.*, has been isolated from various Malaysian **toddy** samples. Yeast counts of fresh toddy range from 1.0 x 10^8 to 3.1x10^9 and bacteria counts range from 1.5 x 10^8 to 3.6 x 10^9 per ml (Merican, 1977).

During fermentation, moulds disappear after 2 or 3 days and aerobic bacteria increase and then rapidly increase with growth of lactic acid bacteria. Lactic acid bacteria increase after 24 hours reaching 2.0x10^8 cells/ml by the 3rd or 4th day; thereafter they decrease to 4.0x10^5 cells/ml. Yeasts increase rapidly to 2.0x10^8 cells/ml after 24 hr. and 8.4 to 10^8 cells/ml in 48 hr. Yeast numbers decrease slightly to 5 to 7x10^8 cells/ml at the end of fermentation (Kim, 1968).

Shin and Cho (1970); Kim (1970); Kim and Lee (1970); Lee and Rhee (1970) isolated *S. cerevisiae, Hansenula anomala, H. subpelliculoso, Pichia polymorpha, Torulopsis sake, T.inconspicua, Bacillus subtilis, B. Megaterium,*

Leuconostoc mesenteroides, and *B. Pumilus. Hensenula spp.* is most important in flavour development.

Kimchi fermentation is due to lactic acid bacteria and yeast naturally present in raw materials (Cheigh and Park, 1994). Carbohydrates, free amino acids and sulfur compounds such as *allyl glucosinolate* and *propenyl-cystein sulfoxide* in **kimchi** are the substrates for the enzymatic reactions and chemical oxidations. Mannose, glucose, fructose, and galactose are the major carbohydrates in **kimchi** and the concentration of carbohydrates decreased during **kimchi** storage (Ha and Other 1989; Cheigh and Park, 1994).

Rehms and Barz (1995) reported the degradation of stachyose, raffinose, melibiose and sucrose by different **tempe** producing *Rizopus* fungi. Worthington and Beuchat (1974) found a decrease of the flatulence-causing sugars stachyose and raffinose during **tempe** fermentation with *Rhizopus delemar*, but not with *R. oligosporus.*

Essential microorganisms in **ragi tape** are moulds of the *Amylomyces rouxii* (Ellis et al., 1976) and yeasts of the *Endomycopsis burtonii* (Ko, 1972). Pichyangkura and Kulprecha (1977) examined 52 samples of mould bran from various parts of Thailand; it is used as a starter in **tape**-like products and for production of alcoholic rice wines. All starters contained *Amylomyces.*

Of 41 yeasts strains isolated from Indonesian **ragi** and **tapé** by Saono (1977) and co-workers, 19 were amylolytic, none were proteolytic, but 14 were lipolytic.

Ko (1977) produced **ragi** containing pure cultures of single microorganisms. He used *A.rauxii* and *Endomycopsis chodati,* each of which was grown separately on cooked rice for 4 days at 30°C. The cultures were then dried and pulverised. When needed, the powders were mixed and inoculated into steamed glutinous rice or casava tubers.

The essential microorganisms for Chinese **laochao** are reported to be *Rizopus Chinensis* (NRRL 3671) and *Endomycopsis sp.* {(NRRL Y 7067) Wong and Hesseltine, 1970)}. In **tape,** ethanol ranges from 3% v/v (Tanuwidjaja, 1972) to as high as 8.5% v/v (Cronk et al., 1977).

A.rouxii and at least one yeast of the *Endomycopsis burtonii* type hydrolyse steamed rice to maltose and glucose and then ferment the sugars to ethanol and organic acids, which provide an attractive flavour and aroma. If yeasts of the *Hansenula* type are present, acids and ethanol are esterified producing highly aromatic **tape.** Fermentation time is 2 or 3 days at 30°C and with continued incubation the product becomes more liquid. Cronk (1974) found that rice lipids were hydrolysed during the **tape ketan** fermentation. Upto 8.5% v/v ethanol serves as a source of calories and probably helps prevent transmission of disease or toxin producing microorganisms in **tape.**

Wong and Jackson (1977) studied Malaysian rice wine.

Uyenco and Gacutan (1977 b) isolated *Endomycopsis fobuligher, Rhodotorula glutinis, Debaromyces Hansenii Candida parapsilosis* and *Trichosporon fennicum, Leuconostoc spp.* homofermentative and hetero fermentative lactic acid bacteria in Phillipine Tapuy. Ethanol content is 5.63% by 48 hours and readers reached 16.01% in 7 days.

In the liquid state fermented products of South Asia, *Amylomyces rouxii* and others play various roles in the brewing Steinkraus (1983).

In Taiwan, there is another series of wines similar to **shao-hsing**, called **hua-tiao** (Wang, 1998; Lin et al., 1975), which uses both wheat **koji** with *Aspergillus oryzae* from solid-state fermentation and submerged rice **koji** with *Rhizopus* species.

In Indian **ruhi** and Indian **madhu,** moulds of genera *Rhizopus* and *Mucor,* Yeast and Lactic acid bacteria hydrolysed starch to sugar and fermented to ethanol. Final ethanol content ranges from 12 to 14 v/v (Dahiya and prabhu, 1977).

Carbohydrate is degraded by the hetero fermentative species by the phosphoketolase pathway (Rainbow, 1965). In contrast to homofermenters, the glycolytic pathway is inactive because *aldolase* and *hexose isomerase* are missing (Wood, 1961). The end product of metabolism includes lactic acid, ethanol, glycerol, acetic acid and carbondioxide.

Homofermenters produce lactic acid but some strains yield a small amount of diacetyl (Barwald et al., 1969). A wide variety of carbon sources will serve, especially maltose (Wood, 1961) but some strains require carbondioxide in addition (Ritter, 1953).

Both homofermentative and heterofermentative strains used a wide range of amino acids, nitrogenous bases and vitamins (Moor and Rainbow, 1955) and therefore growth in beer depends on incomplete uptake of these materials by the yeast. Undesirable effects of the bacteria upon beer include turbidity, acidity and off flavours (Thorwest, 1965). Acetobacteria strains can synthesise their entire complement of nitrogenous compounds from ammonia and suitable carbon fragments (Brown and Rainbow, 1956; Cooksey and Rainbow, 1962). The pH of wort, seeding rate of bacteria and choice of yeast strain are the important factors for the development of the organism in the early stages of fermentation (Case, 1965). Glucose and fructose, sometimes sucrose present in beer, but not maltose are utilized as carbon sources by *Zymomonas anaerobia* (Bexon and Dawes, 1970). Energy is provided by the Entner-Doudoroff pathway and the end products are ethanol and CO_2 (Mc Gill et al., 1965). Minor products such as hydrogen sulfide and acetaldehyde give a highly objectionable odour (Shimwell, 1950).

Many wild types of yeast are strains of *Saccharomyces cerevisiae* or *Saccharomyces carlsbergensis* and cause off-flavour fermentation of dextrins and turbidity of beers (Wiles, 1953). Surveys of wild yeasts have been carried out in Britain on raw materials by Wiles (1953), pitching yeasts by Brady (1958) and in draught beers by Hemmons (1954). According to Wackerbauer and Emeis (1968) all types of *Lactobacilli* can grow in unhopped beer but only a few can grow in the presence of hop resins.

A diverse population of yeasts, including species of *Hanseniaspora, Metschnikowia, candida, pichia* and *Kluveromyces* are often present in the initial stages of most wine fermentations (Fleet, 1998). This non-saccharomyces yeast typically grows for several days before the fermentation is dominated by one or more *Saccharomyces cerevisiac* strains and a concurrent increase in ethanol concentration occurs (Bisson and Kunku, 1993). *Candida* and *Hanseniaspora* species have been shown to persist throughout wine fermentations albeit at a lower level than *S. cerevisiae* strains (Heard and Fleet, 1985; 1988). Lower fermentation temperatures between 10°C and 20°C have been shown to encourage growth and/or persistence of *Kloeckera* and *Candida* species (Hansen et al., 1988) most likely due to increase ethanol tolerance of these yeasts at lower temperatures (Gao and Fleet, 1988). Recently, *Kloeckera* and *Candida* were shown to possess growth rates comparable to that of *S. Cerevisiae* at lower temperatures (10°C) (Charoenchai et al., 1988). Significant growth of non-saccharomyces yeasts early in fermentations has been associated with off character production in wines and sluggish fermentations (Bisson, 1999). Several groups have examined *Candida* species as potential adjuncts for flavour modification of wine (Ciani and Ferraro, 1998; Soden et al., 2000). Yeasts are primarily responsible for the alcoholic fermentation of musts, while many vines undergo another fermentation process mediated by lactic acid bacteria (Fleet et al., 1984). Growth of yeasts on velum surfaces produces important changes in the characteristics of the wine due to the oxidative metabolism of the floor yeasts (Martinez de la ossa et al., 1987). Valcarcel et al. (1987; 1989) Garcia Maiquez (1995) Inigo et al. (1963) Ibeas et al. (1997) Martinez et al. (1995; 1997) studied the various microbiological and aging processes of Sherry wines. Many studies have been conducted to learn more about the non-Saccharomyces yeasts involved in wine making (Amerine et al., 1982). It is generally assumed that the starter suppresses indigenous yeasts. However studies have shown that indigenous yeasts can still participate in the fermentation (Querol et al., 1994; Schütz and Gafner, 1993) or that only 50% implantation was achieved when fermentation was conducted with some commercial strains (Esteve-Zarzoso et al., 2000).

The bacteria are usually combined with yeasts, which play a major role in sour dough leavening by producing ethanol and carbondioxide (Oura et al., 1982).

The filamentous fungus *Aspergillus oryzae* has been used for more than 2000 years for the production of food products in Asia (Cook and Campbell-Platt, 1994) and today it is widely used in the bioprocessing industry for the production of a variety of both native and recombinant enzymes (Christensen, 1994).

Lactic acid fermentation, as with other bacterial fermentation, should be prevented during marc ensilage because it provokes the loss of ethanol, and the formation of compounds, which may be responsible for the off-flavors when present in high concentrations.

Wines are composed of complex mixture of water, phenolic compounds, organic acids and alcohols; and residual sugars are responsible for sensory characteristics of wine. Because of the interfering phenolic compounds, analysis of red wines such as Cynthiana and Syrah is more difficult than with white wines. Standard laboratory analyses for wine incorporate simple physical, chemical and colour measurements used for wine evaluation such as pH titrable acidity, colour and phenolic levels measured by absorbance and transmittance (Iland et al., 1993; Zoecklein et al., 1995). In wine and beer, the yeasts synthesise vitamins, proteins and amino acids and thus, are rich in these compounds besides sugar, acid, phenolics etc. (Garteineau et al., 1979). Wine fermentation kinetics are largely unpredictable, as the composition of grape must, environmental conditions and the robustness of the fermenting yeasts are highly variable. Much industrial wine fermentation can be qualified as "problem fermentation" (Salmon, 1998). Heterocyclic compounds especially sulfur, may play a major role in the aroma of many process products (Maga, 1975). Flavour quality is of major importance to the consumer, but the flavour characteristics of beer appear to deteriorate greatly with time at a rate depending on the beer's composition and storage condition (Lermusieau et al., 1999; Noel et al., 1999; Gijs et al., 2000). The pH factor appears to be a key factor influencing low beer ages. Higher pH at the end of fermentation also favours ester production and yeasts flocculation (Garcia et al., 1994). Esters constitute a major class of flavour compounds in alcohol beverages (Nykänen, 1986). Some physicochemical conditions of grapes such as temperature, pH and sugar concentration are more effective for chemical hydrolysis of terpenol and/or for the transformation and rearrangement of monoterpenes (Gunata et al., 1992). A high pH is unfavourable for a number of important reactions in the brewing process as mashing saccharification will not proceed smoothly, the coagulation of protein at boiling is incomplete and the resultant beer with a high pH are biologically unstable and liable to infection with lactic acid bacteria (de Clerck, 1957). Furthermore, the damaging effect of nitrates on fermentation is diminished (Vogl and Schumann, 1967). Mould growth during steeping is reduced by raising the pH of the steep liquor with lime (de Clerck, 1957).

Aging in wooden barrels permits atmospheric oxygen to pass slowly through its pores, leading to the gentle oxidation of certain components of the wines, a reduction in astringency, and changes in colour and taste (Noble, 1990; Vivas, 2000). At the same time certain components are extracted from the wood into the wines, giving the wines their characteristic aromas and taste (Peyron et al., 1994; Chatonnet and Dubordieu, 1998). Phenolic compounds are mainly responsible for these characteristics (Glories, 1990) although they must evolve suitably during the aging process if a quality wine is to be produced (Puech et al., 1988). The phenolic levels recorded were within the range considered acceptable for the aging process to be used (Ruiz, 1994). The phenolic content of the wines was evaluated by the total polyphenols index (Riberéau and Peynand, 1971).

Some authors have related the Vanilla aroma with the cir-whisky lactone content of red wines (Spillmen et al., 1997). Zhou et al. (2000) reported Hexanal as an unconsiderable flavor-volatile generated during the fermentation of cucumber. Benzaldehyde noted for imparting bitter almond and cherry flavors, has been identified as a secondary metabolite from microbial fermentation by *Pseudomonas putida* (Romero, 1992). Yokotsuka (1981) reported that *Torulopsis etchellsii* and *T.versatilis* contribute to the flavour in **shoyu**. Yoshida's (1968) recent studies emphasise the importance of having the proper amount of amino acids in wort for achieving a satisfactory fermentation. The composition thereof is dependent both on the quality of the raw materials and mashing procedure (Enari et al., 1970).

Hops in addition to its antiseptic value (Wackerbaver and Emeis, 1968) contribute to the foam-stabilizing properties of wort and are important for achieving satisfactory yeast head production in ale- type beer (Dixon and Kirsop, 1969).

Yeast requires sulfur for the production of proteins coenzymes, vitamins etc, and takes up organic sulfur from wort, chiefly as methionine and inorganic sulfur in the form of sulfate (Maw, 1965).

Acidity is one of the most important organoleptic parameters in wine and is mainly due to the presence of weak organic acids. All the acids in wine are already present in the grapes, but very small quantities of organic acids such as succinic acids are produced during alcoholic fermentation (Ramon-Portugal et al., 1999). The composition and concentration of each acid is essential for the quality of the final wine.

Tartaric and Malic acids are the main acids in grapes (upto 90%). Succinic acids citric, pyruvic, lactic and gluconic acid are in lcw molar concentrations and contribute little to the titrable acidity and pH of wine. Tartaric acid concentration remains practically constant but malic acid decreases during grape ripening (Riberéau-Gayon et al., 2000).

Organic acids are important not just because of their organoleptic contribution but they also play an important biotechnological role in industrial fermentations. These acids are in equilibrium with their salts and they act as a buffer and thus maintain the pH of wines in the range from 2.9 to 4 (Dartiguenave et al., 2000).

The acidic components of fruit juices and wines impart important properties to these foods being prominent in flavour, processing and preservation. These acids have been extensively studied in the past (Amerine and Cruess, 1960; Hulme, 1970; Tressler and Joslyn, 1961).

Brunelle et al. (1967) and again Martin et al. (1971) found wine to contain the acids succinic, tartaric, malic and citric. Fernandez-Flores et al. (1970) reported that most of the above acids are present in common fruits as well as significant amounts of furmaric and glycolic acids are also found. Markakis and Kallifidas (1971) investigated the organic acids of Concord grape wine and found a variety of common acidic compounds plus other minor types such as galacturonic and lactic acids.

Phenolics in wine on the other hand are important in furnishing the colour of wine, producing the astringent taste, and offering a reservoir for oxygen reduction (Singleton and Esau, 1969). Variety, processing practices and storage, among other factors affect the phenolic composition of grape juices and wines (Amerine and Ough, 1980).

Previous work has shown that the tartaric esters of caffeic and coumaric acids are the major cinnamics (Riberéau-Gayon, 1965; Singleton et al., 1978; Baranowski and Nagel, 1981). Procyanidins built on epicatechin and catechin units make up the procyanidin fraction (Lea et al., 1979) of grape juices and wines. Flavonol derivatives (Cheynier and Rigaud, 1986) and hydroxybenzoic acids (Singleton and Trousdale, 1983) also occur.

Beer contains many phenols - the greater part of which comes from the malt and the remaining portion from the hops (Goupy et al., 1999). Phenolic acids can be found in germinated barley as free form or bound to the cell walls. They contribute to the antioxidant activity in beer (Maillard, 1995). Phenols in beer are present in both monomeric and polymeric forms.

Recently, Montanri et al. (1999) determined 16 phenolic acids in beer by coulometric array detection using a binary gradient of 0.1M-phosphate buffer. Xanthohumol a prenylated chalcone is the principal flavonoid in hop, constituting 80 - 90% of the total flavonoids. Minor amounts of other prenylchalcones and prenylnaringenins have also been reported (Stevens et al., 1997). Of these constituents, Xanthohumol and 6-prenylnaringenin have been found to have antifungal activities (Mizobuchi and Sato, 1984).

A variety of tree materials are known to be remarkable sources of phenolic compounds. Several phenolics have been isolated and identified from conifer needles (Lundgren, 1987; Strack et al., 1989, bark of birch, spruce and pine (Pan, 1995); and from white birch and silver birch leaves (Ossipov et al., 1996). Pyenogenol (procyanidin extract of the bark of *pinus maritima*) is probably the most studied phenolic tree extract. It has been shown to scavenge free radicals, including hydroxyl and superoxide anions (Noda et al., 1997), and it may have beneficial effects in preventing antheroselerosis and other age-related diseases (Fitzpatrict et al., 1998).

Phenols of three different types were found in silver birch inner bark (phloem); arylbutanoid glucosides, benzoic acid derivates and catechins, suberin and the triterpenes betulin (10 - 30% of the outer bark), betulinic aldehyde, and acid lupeol, erythdiol, oleanic acid and several triterpene caffeates were identified in outer bark of silver birch (Pan, 1995). The latter compounds have previously been shown to have antioxidant (Takagi and Iida, 1980) and anti-inflammatory properties (Otsuka et al., 1981). Isorhapontin was identified as the most efficient fungi static compound in phenolic extracts of Norway spruce. Especially high isorhapontin concentrations were found in the outer root bark, stem bark and fine root extracts (Beyer et al., 1993).

One of the most important phytochemical groups is phenolic; plant phenolics are antioxidants in many food systems (Mau et al., 2001; Kahkonen et al., 1999). Phenolics prevent food deterioration (Shahidi and Naczk,1995). Phenolics show both antioxidant activity and antimuctagenicity and are functional food components possessing health benefits or being able to prevent disease in human beings (Yen and Chen,1995). Plant phenolics are present in free, ester and insoluble bound forms (Shahidi and Naczk, 1995; Krygier et al., 1982). Free and esterified phenolics are soluble in methanol and found to be the predominant phenolics in some plant (Krygier et al., 1982). Hydrolysis will convert the esterified phenolics to simple phenolic acids. There are also some reports about a heat-stable antioxident of phenolic origins carried by the superoxide dismutase isolated from dried peas (Nice et al., 1995).

Phenolic acids such as hydroxybenzoic acids and hydroxycinamic acids are antioxidant compounds in fruits and vegetables (Bagchi et al., 1993). Jackson et al. (1967; 1968) isolated a new series of biflavanones from *Garcinia buchananii* Baker and *G. engeniifolia* Wall. Eldridge (1982) reported on the use of HPLC for quantitating phenolic compounds in soybeans. Some biflavonoids which have also been isolated from *Allamblackia floribuda* Oviver (Locksley and Murray, 1971), belong to the same sub-family of *Clusiodeae,* as *garcinia;* the presence of these biflavonoids extracts seems to be characteristic of this sub family. Derivatives of the five common hydroxy benzoic acids salicyiic (2-OH), 4-hydroxybenzoic, vanillic (3-OMe, 4OH), protcatechnic (3,4-di OH), and syringic acids (3,5-0Me, 4-OH) are widespread in the plant kingdom (Billek and Kindl, 1962; Harborne and Williams, 1971) even if their concentration is usually low as compared to the ubiquitous hydroxycinnamic acid derivatives. Hydroxybenzoic acid derivatives have been found only in a few plant species (Herrmann, 1978). Due to the behaviour of hydroxybenzoic acid derivatives on hydrolysis it has been assumed that these acids mainly occur as glucosides rather than as glucose esters (Durkee, 1977; Senter et al., 1983).

Isolation and structure elucidation of syringic acid 4-B-D-glucoside has been successful from *Anodendron affine* (Shima et al., 1971) and *Eukianthus nudipes* (Ogawa and Ogihara, 1976). Protcatechuic acid 3-b-D-glucoside was found in Pyrus (Challice and Williams, 1968).

Gallic acid is different from other hydroxybenzoic acids, in being esterfied to glucose in the form of hydrolysable tannin (Herrmann, 1978).

Members of the *Apiaceae* were investigated to determine the content of 4-hydroxybenzoic acid glucoside (Dirk and Herrmann, 1984) and soft fruits were examined for the presence of 4-hydroxybenzoic, protocatechuic and gallic acid glucoside (Schuster and Herrmann, 1985; Schuster et al., 1986). Manniflavanone, A new 3, 8-linked flavanone dimer from the stem bark of *Garcinia mannii* were determined (Crichton and Watermann, 1979).

Santos and Linardi (2001) studied the phenol degradation by yeast isolated from industrial effluents.

Piga et al. (2003) studied the influence of drying parameters on polyphenols and antioxidant activity.

Improved high-performance liquid chromatographic analysis of phenolic acids and isoflavonoids from soybean protein products were determined by Seo and Morr (1984).

Shahrzad (1996) determined some pharmacologically active phenolic acids in juices by High Performance Liquid Chromatography.

Phenolic acids present in berries are hydroxylated derivatives of benzoic acid and cinnamic acid (Macheix et al., 1990).

Caffeic, Chlorogenic, ferulic, Sinapic and p-coumaric acids appear to be more active antioxidants than hydroxy derivatives of benzoic acid such as p-hydroxybenzoic vanillic and syringic acids (Larson, 1988).

Ferulic acid from Brewer's Spent Grain (BSG) (Hernanz et al., 2001) exhibit a number of potential commercial applications, like food flavour precursor (Graf, 1992).

Caffeic acid is one of many phenolics considered to be an important part of the general defence mechanism of plants against infection and predation (Faulds and Williamson, 1999; Friend, 1985). The production of proteases by *Rhizopus oligosporus* decreased as the concentration of substrate increased (Wang et al., 1969).

Recently, Tsujisaka et al. (1972) found that *Aspergillus niger* produced more lipase on solid medium containing wheat bran and calcium carbonate than in liquid media. Soybeans were not as good a substrate for protease production as was wheat bran.

Wang et al. (1974) reported the production of acid protease by fungi used in soybean food fermentation.

Baking quality of flour is influence by fungal a-amylase supplementation on mylolytic activity (Curiæ et al., 2002).

Studies revealed that *R. oligosporus* (Wang and Hesseltine, 1965), *M. dispersus* (Wang, 1967) and *Aspergillus elegans* (unpublished data) would produce acid type proteases, whereas most fungal proteases have a pH optimum around neutral or alkaline. Weimberg and Oxfon (1966) observed that acid phosphatase and invertase of *S. cerevisiae* and *S. fragilis* were located outside the cell wall but were bound to the cell.

Penillinase of *Bacillus subtilis* liberated from cell or enveloped by an enzymatic process was discussed by Kushner and Pollock (1961). Mallamy and Horecker (1961) found that when cells of *Escherichia coli* were converted to spheroplasts with ethylenedia minetetra acetate and lysozyme, the surface located alkaline phosphatase could be released into the medium. This method was used to liberate other surface located enzymes of *E. coli* such as ribonuclease, acid phosphatase, cylic phosphodiesterase and 5'-nucleotidase (Neu and Heppel, 1964).

The protase produced by *M. dispersus* NRRL 3103 was bound to the mycelial surface (Wang, 1967). Addition of sodium chloride or other ionizable salts in the growth medium for this fungus increased the enzyme in the culture filtrate.

Yabuki et al. (1977) and Erratt et al. (1984) reported that á-amylase production was induced by carbohydrates containing á-1, 4 glucosidic bonds, (maltose, maltodextrins or starch). It was stated that glucose effectively repressed á-amylase production. Glucose was also reported to repress á-amylase production in the closely related fungus *Aspergillus awamori* (Bhella and Altosaar, 1987). Lachmund et al. (1993) found that starch and maltose strongly increased á-amylase productivity in *A. oxyzae,* whereas glucose led to very low productivities (approximately 10% of the level obtained with starch). Since addition of glucose to a starch medium gave the same productivity as for starch only. Lachmund et al. (1993) concluded that glucose did not cause any carbon catabolite repression of the applied strain of *A. oryzae*. Based on chemostat experiments with *A. oryzae,* Carlsun et al. (1966 a) found a significant repression of á-amylase production by glucose even at 10mg/l glucose. Møkeberg et al. (1995) reported that maltose induces both the synthesis and the secretion of á-amylase. Pedersen and Nielson (2000) studied the influence of nitrogen sources on á-amylase production by *A. oxyzae*. Rice bran and paddy husk were used as alternative substrates (Pandey et al., 1994 a, b) and/or support (Arasaratnam et al., 1977) for glucoamylase production. Inclusion of paddy husk helped to yield more glucoamylase production than rice bran by efficiently utilizing the substrate (Arasaratnam et al., 1997).

Experiments have been performed to study the effects of ã-irradiation on various aspects of wheat (Koksel et al., 1996), rice (Sabularse et al., 1991) and maize (Sokhey, 1993) quality. Much research has been done on the effects of ã-irradiation on starch properties from various sources (Bao et al., 2001; Sokhey and Chinnaswamy, 1993; Rayas-Duarte and Rupnow, 1993). Free radicals produced by ã-irradiation modify the amount and structure of starch fractions (Raffi et al., 1998). Increased radiation dose is invariably associated with lower viscosity and higher water solubility and acidity of starches (Bao et al., 2001; Mac Arthur and D' Appolonia, 1984; Bachman et al., 1997; Sokhey et al., 1993). The chain length of amylose and amylopectin shows a progressive reduction as the irradiation dose is raised (Sokhey and Chinnaswamy, 1993; Roushed et al., 1983; Rayas-Duarte and Rupnow, 1993). Bao et al. (2001) showed that increasing levels of ã-irradiation significantly reduced rice flour viscosity and modified the pasting profile shape. Raffi et al. (1980) and Michel et al. (1980) did both theoretical and experimental studies of the depolymerization of starch under the combined action of protons and ã-radiation. Kang et al. (1999) tried to modify maize starter by ã-radiation and pH adjustment and found that the combination considerably affected viscosity.

Depolymerization of starches is higher in the presence of acid. Acid such as HCl and H_2SO_4 cause scission of the glucosidic linkages, thereby altering the structure and properties of the native starch (Hover, 2000).

Acid modification of Oat starch granules caused a marked reduction in viscosity, degradation of amylopectin to smaller molecules, and formation of more elastic gel structure based on amylose (Virtanen et al., 1993).

Ethanol is an important industrial chemical with emerging potential as a biofuel to replace vanishing fossil fuels (Von sivers et al., 1994; Wheals et al., 1999).

By modifying nutritional conditions, it is possible to increase ethanol production by traditional fed-batch process with *S. cerevisiae* up to 11°C Gay Lussac (°GL) (85.8g I^{-1}) in 125 hours (Panchal and Stewart, 1981).

Enhanced ethanol productivity has been demonstrated during glucose fermentation with *S. cerevisiae* (Cheryan and Mehaia, 1984) and during xylose fermentation with *Pachysolen tannophilus* (Chung et al., 1986).

Casey and Ingledew (1985); Van Uden (1985); D' Amore and Stewart (1987), Jones (1989) comprehensively reviewed the critical concentration of ethanol at which yeast is grown and influenced by several factors.

3

Meterials and Methods

3.1 Source of raw material

Two varieies of *Albizia, Albizia kalkora* Pain (Syn.*Mimosa kalkora* Roxb) and *Albizia myriophylla* Benth.commonly used by brewers in rice fermentation were collected from the forest of Manipur. Fresh bark was extracted from the tree or twig and sun dried. Dry bark was kept in the oven at 100°C for 2 hours before crushing to remove the moisture. Dry bark was crushed into powder by mechanical grinder and stored at ambient temperature for the experiment.

Different rice varieties and cultivars commonly cultivated in the state (Manipur) were collected. RCM-5, RCM-9 and RCM-10 were obtained from Indian Council of Agricultural and Research (ICAR), Imphal, KD-2-6-3, IR-64, CAUR-3 and *Tamphaphou* were obtained from Central Agriculture University (CAU), Imphal. **Mpangban, Tunanapnang, Napdai, Banglai, Ritiang, Nappuak, Sangbu, De se** and **Shazei** were obtained from Tamenglong District, Manipur. Mixi-grinder was used to dehusk the rice and grains were separated from the husk and stored at ambient temperature for the experimental purpose.

3.2 Determination of moisture

The moisture content was determined by the method of

A.O.A.C (1970). Samples were accurately weighed and dried in and oven at 100°C (±2°C) until constant weight was obtained. The estimation was done in triplicate and the mean values of both were recorded to calculate the moisture content using the following relationship.

$$\text{Moisture content (g/100) sample)} = \frac{\text{Initial weight - final weight.}}{\text{Weight of the sample}}$$

3.3 Determination of Ash

The Ash content was determined as described in the A.O.A.C method (1970). For this purpose, a weighed amount of moisture-free powdered sample (1 gm) was taken in a crucible, charred and finally ignited at 550°C for 6 hours in a muffle furnace. The estimation was done in triplicate and the mean was expressed as mg/g of ash content in moisture-free sample.

$$\text{Ash content (g/100g sample)} = \frac{\text{Weight of the ash}}{\text{Weight of the sample}} \times 100$$

3.4 Determination of total soluble sugar

Total soluble sugar was estimated by using anthrone reagent (Dubois et al., 1951). Here, anthrone reagent was used in colour development resulting in blue-green solution. In this, 50 mg of sample was crushed in 10 ml of 80% ethanol. The content was centrifuged for 10 minutes at 4000 rpm. Then the supernatant is made up to 10 ml.

Further, 10 ml of the above extract and 4 ml of anthrone solution was added and heated for 10 minutes in boiling water bath and was allowed to cool at room temperature. The intensity of blue-green colour was measured in systronic-UV spectrophotometer at 625nm against a reagent blank. The sugar content was estimated from a standard curve prepared with known concentration of glucose.

3.5 Determination of Reducing sugar

Reducing sugar in rice grain and bark was determined by following Nelson's modification of Somogyi's (Nelson, 1944) method. In this, 50 mg of sample was crushed in 10 ml of distilled water with a pinch of activated charcoal. The content was centrifuged for 10 minutes at 5000 rpm. Then the supernatant is made up to 10 ml in a measuring cylinder.

One ml of this extract was added in a test tube containing 2 ml solution of Somogyi reagent (4 ml of reagent A + 1 ml of reagent B). The mixture was incubated at 100°C for 10 minutes in water bath. After cooling the contents, 4 ml of arsenomolybdate reagent was added. The intensity of blue-green colour was read at 640 nm in a systronic-UV spectrophotometer. The values were calculated by a standard curve prepared from glucose.

3.6 Determination of non-reducing sugar

The non-reducing sugar content was calculated by subtracting the percentage of reducing sugar from the percentage of total soluble sugar.

3.7 Determination of total free amino acid

The estimation of total free amino acid was done by the method of Yemn and Cocking (1955). In this, 50 mg of dry powdered samples were homogenised in 10 ml of 50% ethanol with a pinch of activated charcoal. The contents were centrifuged at 1000 rpm for 10 minutes and the free amino acids were collected in the form of clear supernatant.

To 1 ml of supernatant, 25ml of 2% ninhydrin (w/v in Isopropyl alcohol) and 2.5 ml of 0.1M acetate buffer (ph 5.5) were added. The mixture was heated on boiling water for 30 minutes and after cooling, aqueous isopropyl alcohol (1:1) was added to make up the volume to 10 ml. The colour intensity of the violet complex was measured at 570 nm as absorbance. The amount of total free amino acids was calculated with the help of a calibrated curve prepared from glycine and was expressed as mg amino acids powdered per gm dry weight of the sample.

3.8 Determination of starch

The starch content in the sample was estimated by the method described by Chopra and Konwar (1976). The sugar free residue left in the centrifuge tube after alcoholic extraction (hot 80 % ethanol) was dried at 70°C before the extraction of starch.

The starch was extracted with 52% perchloric acid by stirring for 30 minutes. This was then centrifuged and supernatant was collected in a volumetric flask (50 ml). The process was repeated for 2 to 3 times and the volume of the combined extract was made upto the mark with 50 ml perchloric acid.

The starch content in the extract was determined by using anthrone as described for sugar estimation. The value obtained from standard curve was multiplied by a factor of 0.9 to get the starch content. The estimation was carried out in triplicate and the mean was expressed as starch in per-100 gm moisture-free sample.

3.9 Determination of amylose content

Amylose content can be determined by the method described by Swaubhagya and Bhattacharya (1979).

Defatted moisture free powder sample was taken in stoppered conical flask. One milliliter of dissolved alcohol was added to wet the powder. Ten ml of 1N NaOH was then gently added by the side and kept overnight, and both were heated the next day in boiling water for 2 to 3 minutes. After cooling, the volume was made up to 100 units with distilled water.

Five milliliter of the above dispension was transferred into a 100 ml volumetric flask. Fifty ml of water was added to it followed by addition of 1 ml acetic acid and 2 ml of iodine (0.2 %) solution. The volume was made upto the mark with distilled water and the pH of this solution was 4.5.

One milliliter of standard amylose solution was taken and treated by using the same procedure. A reagent blank was also similarly prepared except for the addition of amylose extract on the standard amylose solution. The intensity of the colour developed was measured in systronic-UV spectrophotometer at 630 nm after 30 minutes against the reagent blank. Amylose content was calculated using the following relationship.

Calculation

Amylose is expressed as percentage of starch according to the formula-

$$\text{Amylose content} = \frac{R}{A} \times \frac{a}{\gamma} \times \frac{1}{5} \times 100$$

Where, R = 630 nm reading for sample dispension
A = 630 nm reading for standard amylose solution
a = Amount of standard amylose taken
γ = Amount of sample taken

3.9.1 Determination of amylopectin content

The amylopectin content was determined by subtracting the percentage of amylose from 100 on moisture-free basis.

3.10 Estimation of phosphate buffer solution protein

The phosphate buffer solution protein was estimated by Lowry et al. (1951) method, which evolves the following steps.

Reagents

a) Reagent "a" = 2% sodium carbonate in 0.1N.NaOH

b) Reagent "b" = 1% copper sulphate

c) Reagent "c" = 2% Sodium potassium

d) Reagent "d" = a + b + c (10+0.1+0.1)

e) Reagent "e" = Folin-Ciocalteau phenol reagent.

f) 10% T.C.A = 10 mg of T.C.A in 100 ml of the solution.

g) 95% ethanol = 95 ml of dehydrated alchohol with 5 ml of D.W.

Here, 500 mg of fresh bark sample and 250 mg of rice powder were crushed to 10 ml cold phosphate buffer (PH 7.5, 0.1 M) using a chilled motar and pestle. The homogenate was filtered through four layers of cheesecloth. The filtrate was centrifuged at 5000 rpm for 15 minutes in a centrifuge. To save the loss of protein due to thermal denaturation, ice pieces in a metallic tubes and the small glass centrifuge were kept in a centre. This device was a temporary alternative to the refrigerated centrifuge. The supernatant was poured into a graduated test tube and its volume was raised to a certain level (10 ml) with the same buffer. The pellet was discarded. The supernatant contains soluble proteins. An amount of 1 ml from this supernatant was taken in another centrifuge tube and 1 ml of trichloroacetic acid (TCA) solution (10%, W/V in distilled water) was added to it. The precipitate (whitish in colour) was observed immediately. Precipitation was completed in 5 minutes. The precipitation was centrifuged and after centrifugation the supernatant was discarded. A little amount (5 ml) of the ethyl alcohol (95%) was added into the protein precipitate so as to remove the excess TCA from the surface of the protein.

The washed precipitation was mixed with a freshly prepared reagent "d" at room temperate (30 ± 2°C) and kept for 20 minutes with occasional shaking. Later on, 0.1 ml of the diluted (1:2 with distilled water) reagent "e" (Folin Cioceteau's reagent) was added and allowed to stand for 30 minutes at room temperature. A bottle green to navy blue colour appeared. Then its absorbance was read at 750 nm by using a spectrophotometer. The standard curve for protein is prepared by a standard protein bovine serum albumin (BSA).

3.11 Determination of total phenol

Phenol was estimated by Folin Ciocalteau method conducted by Bray and Thorpe (1954).

500 mg of fresh bark and 50 mg of dry bark samples were homogenised with 80% ethanol and centrifuged at 5000 rpm for 15 minutes. The supernatant was collected and the final volume was made up to 10 ml with the same alcohol (80%).

One ml of alcohol extracted was mixed with 1 ml of Folin Ciocalteau reagent and 2 ml of 20% sodium carbonate solution. The whole mixture was shaken and then heated in a hot water bath for 1 minute. Then the volume was made up to 5 ml with distilled water after cooling and the absorbance was measured at 650 nm in a Systronic-UV spectrophotometer. A reagent blank with sample was also prepared. The amount of total phenol was calculated from a standard curve made from different concentrations of catechol (0.1mg/ml).

3.12 Determination of organic acids

Determination of organic acid was done by modified method of Ryan (1972) and Picha (1985). All solvents (analytical reagent grade) were redistilled before use.

3.12.1 Extraction of organic acid

15 gm bark powder of *A. myriophylla* and *A. kalkora* were refluxed in 200 ml distilled water at 80°C for 1 hour and filtered through Whatman No.42 paper; then the filtrate was freeze dried. The dried powder was dissolved in 30 ml methanol and centrifuged at 12000 rpm for 16 minutes (x 3 times). The aqueous methanol extract (supernatant) was

concentrated under N_2 gas. The concentrate was treated with 1N NaOH (pH 10) and extracted with $CHCL_3$ (Chloroform). Aqueous NaOH was again treated with IN HCL (pH 2) and freeze dried. The powder was eluted with 1 ml MeOH and subjected to TLC analysis.

3.12.2 Thin Layer Chromatography

Thin Layer Chromatographic (TLC) method was employed for the separation of organic acid. Silica Gel-G (E-Merck) Bombay, India, was used as absorbent for the process. Ten gm of silica powder mixed with 20 ml of distilled water was shaken vigorously and spread on a glass plate of 20cm x 20 cm at a thickness of 0.25mm/0.5mm with a 'Desaga' spreader. The coated plates were left to dry and set overnight at ambient temperature then activated at 110°C for 1 hour. Plates were cooled and stored in desiccator until used.

Solvents

Separation of organic acid was carried out in the solvent system of Toluene Ethanol and Formic acid (5:4:1). Twenty ml of this solvent was kept in the tank and allowed it to equilibrate for 30 minutes.

3.12.3 Application of sample

10 to 15 ìl of the methanol extract was applied as a band on the activated silica gel TLC plate using microtip and pipette in regular intervals. The plate was developed with the proposed solvent in a chromatographic chamber and subsequently dried at ambient temperature. Authentic organic acid (from Merck) was used as a standard.

Spray reagent (0.04% w/v bromophenol blue, 0.5% w/v Sodium acetate, dissolved in 95% ethanol and filtered with Whatman No. 4 paper) was taken in spray bottle.

For qualitative identification of the organic acid, the developed plate was sprayed with the proposed spray reagent and dried to reveal/detect the acids.

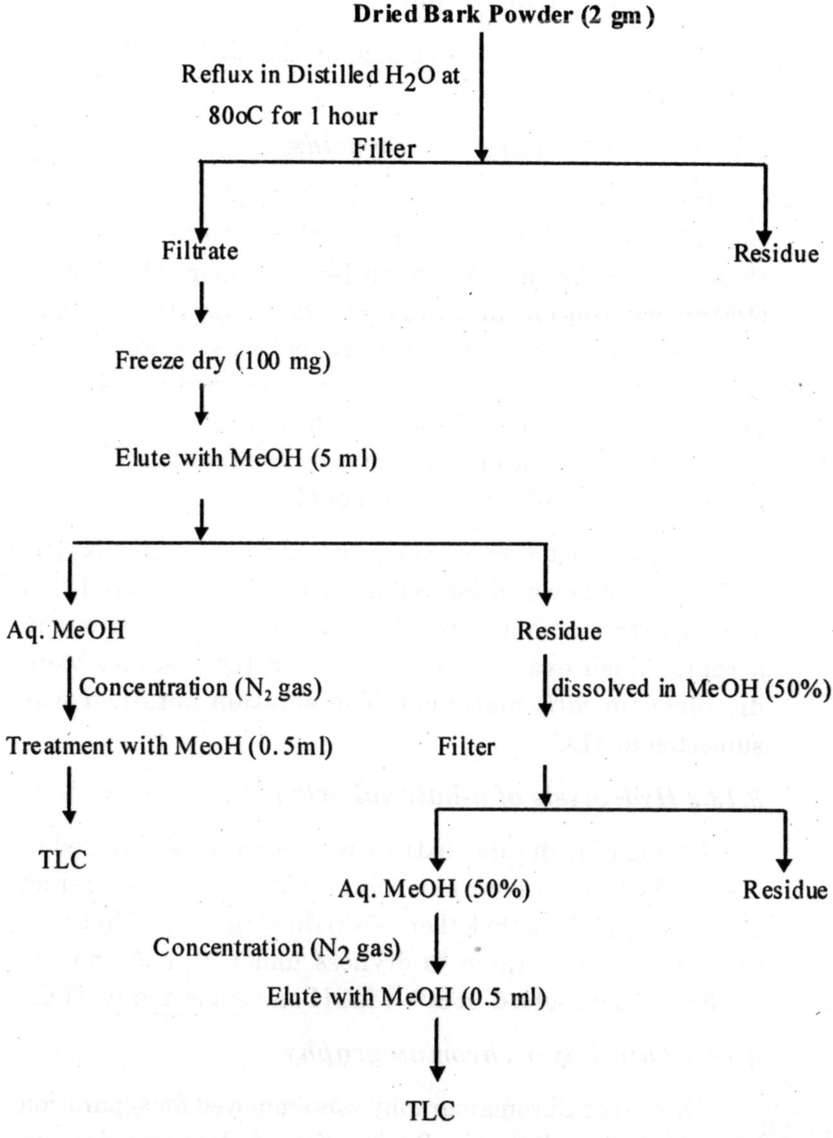

Fig. 1: Flowchart of organic acid extraction of *Albizia* bark

3.13 Determination of phenolic acid

Phenolic acid was determined by the method of Variyar et al. (1988).

3.13.1 Extraction of phenolic acids

50g bark powder of *A. myriophylla* and *A. kalkora* was refluxed in 80% methanol (250 ml) at 80°C for 2 hours and then filtered through Whatman No. 42 paper. The filtrate (160 ml) was concentrated in a rotary flash evaporator, Buchii rotavapor, at 35°C until the methanol was removed. The aqueous residue was freeze-dried and the dried powder was dissolved in 50 ml distilled water. The aqueous solution was extracted three times in each case by Diethylether (50.x 3), Ethylacetate (50x3) and n-Butanol (50 x 3).

Ether extract was evaporated to dryness under N_2 gas and the residue was dissolved in $CHCl_3$ (Chloroform). Ethyl acetate extract and n-Butanol extracts were concentrated in a rotary flash evaporator at 35°C and the residues were dissolved in 80% methanol. The solution obtained was subjected to TLC.

3.13.2 Hydrolysis of n-butanol extract

500 mg of n-Butanol extract was hydrolysed/refluxed in 30ml 1N HCl for 1 hour in a boiling water bath. The extract was treated with diethylether (50x3) three times. Diethylether extract was evaporated to dryness under N_2 gas and the residue was dissolved in 80% MeOH and subjected to TLC.

3.13.3 Thin Layer chromatography

Thin layer Chromatography was employed for separation of various phenolic acids. Preparation of plate was done as describe in 3.12.2. The solvent system Dichloromethane Toluene and Formic acid (5:4:1) was used.

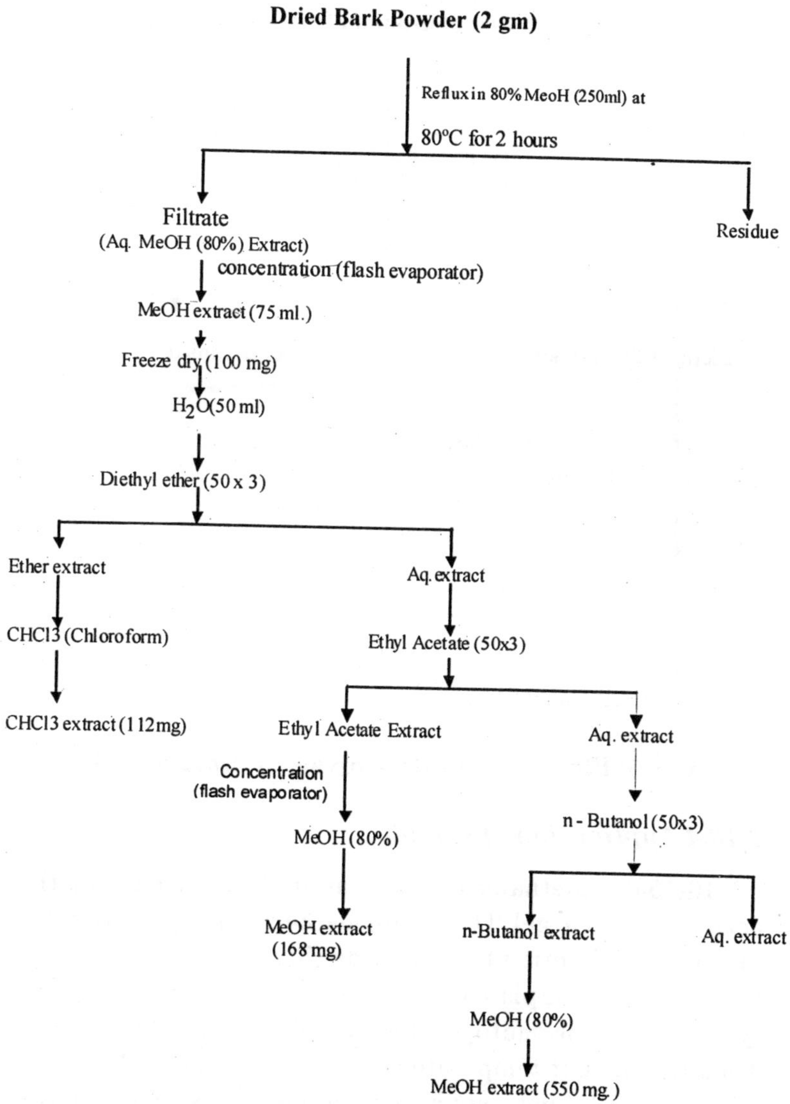

Fig. 2: Flowchart of Phenolic acid Extraction of Albizia bark

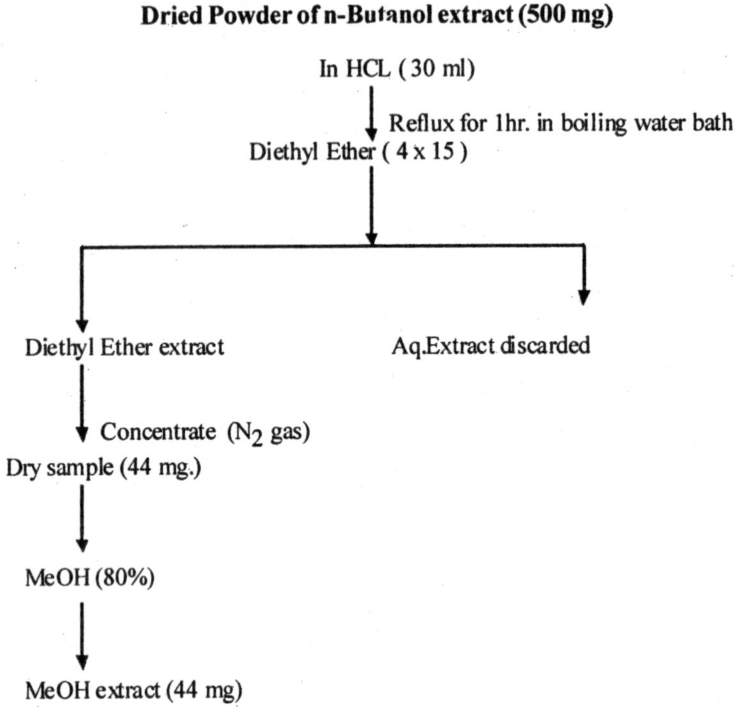

Fig. 3: Flowchart of Hydrolysis of *Albizia* bark

3.13.4 Application of sample

10-15 ìl of methanol extract was applied as a band on the activated silica gel TCL plate using microtip in regular interval. Authentic phenolic acids (Sigma USA) were used as a standard. The plate was developed in the proposed solvent system in a chromatographic chamber and subsequently dried at ambient temperature. Spraying reagent (1% ferric chloride and 1% potassium ferricyanide) was sprayed on the developed plate and left to dry. After detecting the spot the RF values were calculated according to the formula:

{RF = Distance traveled by the compound / Distance traveled by the solvent}

$$RF = \frac{\text{Distance traveled by the compound}}{\text{Distance traveled by the solvent.}}$$

Phenolic acids from the sample were detected by comparing the RF value of the standard.

3.13.5 Analysis of Penolic Acids by HPLC

3.13.5.1. Sample preparation

30g each of ether extract of *A.myriophilla* and *A. kalkora* (hydrolysed and unhydrolysed) were dissolved in Methanol HPLC grade (E.Merck India Ltd. Bombay), to make 0.1% solution and filtered through 0.45 ì filters from Millipore Filter Corporarion, USA. The prepared sample was kept for HPLC analysis.

A Pharmacia LKB HPLC system equipped with model-2248 pumps, a LCC-2252 gradient controller, a VWM-2141 double wave-length UV detector set at 292 nm and a Rheodyne model 7125 injector with 200 ìl sample loop were used; peaks were plotted by a Plotter.

All solvents (analytical reagent grade) were from E Merck (India Ltd), Bombay, India. Solvents were redistilled before used. HPLC grade Methanol, Milli Q water and Acetic acid solution filtered by 0.45-µ filters (from Millipore filter Corporation) USA were used. Standard phenolic acids were from sigma chemical company (USA).

3.13.5.2 Qualitative analysis

Qualitative analysis was carried out on a Mino RPCS 5/20, C_2/C_{18} analytical column (Pharmacia LKB Biotechnology Uppsala, Sweden) 4.6 x 200mm ID. The samples were eluted with 2% acetic acid as solvent A and methanol as solvent B using linear gradient elution from 0% to 100% B in A over a period of 30 minutes at a flow rate 1.2 ml min^{-1}.

Samples (100 ìl each) of unhydrolysed and hydrolysed preparations were used for this assay. Peaks were identified by comparing their retention time with that of authentic standards injected separately under the same conditions.

3.14. Isolation and characterisation of microorganism from *Albizia* bark

Isolation and Characterisation of microorganism was done by the modified method of Burnett et al. (1983) and Buchanan et al. (1974)

3.14.1. Preparation of LB (Luria Bertani) and YPDA (Yeast-Extract peptone Dextrose Agar) media

L.B - Dissolved 2% of Luria Bertani, 2% Agar in a distilled water and autoclaved for 20 minutes at 121°C, bring the media into 45-50°C and poured into sterile petriplates and allowed it to solidify.

YPDA - 1% yeast extract, 2% peptone, 2% dextrose and 2% Agar were prepared in distilled water and adjust pH to 5.7 by 1N HCL. Media plates were prepared as described above. The solidified plates were kept air tight and uncontaminated for future use. These media were used for the isolation of bacteria, fungi and yeast.

3.14.2. Isolation and identification of microorganisms

One gm of bark powder was suspended in sterile saline (0.85%) and serially diluted using the same.

Serial Dilution

Eight sterile eppendorf tubes containing 0.9 ml (900 ìl) sterile saline were taken and labeled as number 1, 2, 3 and up to 8.

One gm of bark powder was suspended in 1 ml of sterile saline (0.85%) in the sterile eppendorf tube and vortex. Then, 0.1 ml (100 ìl) of the sample suspension was aseptically transfer to another blank tube number 2 through a sterile microtip by pipette. Thus serial dilution was continued up to 8th tube by using fresh sterile tip for each dilution.

100 μ l of each dilution was transferred into a sterile media plates, plated and incubated at ambient temperature for 48 hours.

Well-isolated, putative, yeasts, moulds and bacteria colonies were selected, re-streaked on fresh plates. Isolated single colonies were preserved for identification and fermentation experiments:

3.14.3. Determination of total microbial load

One gm of bark powder was suspended in 1 ml sterile saline (0.85) and serially diluted up to the 8th tube and plated on the sterile media plates as described above. At the end of incubation, the total microbial load (cfu) colony-forming unit was determined taking dilution factor into consideration by the following formula.

{Original cell density = No.of colonies/ ml plate x 1 / dilution factors}

$$\text{Original cell density} \; \frac{\text{No. of colonies}}{\text{ml plate}} \; x. \; \frac{1}{\text{dilution factors}}$$

3.15. Determination of fermentative abilities on individual carbohydrates by five isolated microorganisms

Five ml of YPDA (Yeast extract 1%, Peptone 2%, Dextrose 2% and Agar 2%) (pH 5.7) was distributed in a test tube, placed Durham's tube invertedly inside, plugged with cotton and sterilised for 15 min at 15 p.s.i. (121°C).

A solution of 2% glucose, 2% galactose, 2% sucrose, 2% maltose and 2% lactose was separately prepared and filtered through a membrane with the help of sterile syringe and holder.

Test tubes in a set of five of five (5x5) were arranged and the five isolated microorganisms were aseptically inoculated. Fermentation activities of the five isolates on five different carbohydrates were observed.

3.16. Fermentation

3.16.1 Traditional method of Hamei preparation

Chamang rice (unpolished) was soaked in water for about half an hour (then the water was drained out), spread and dried for about 15 minutes to remove the excess water. Presoaked rice was crushed/pounded into powder. In the mean time the dried bark of **yangli** (*Albizia* plant) was cut, chopped and soaked with small quantity of water until a brown colour was developed. The two materials were mixed in appropriate proportion and water was added to the mixture to make a thick paste. A small round cake (***hamei***) was prepared according to the convenience of the practitioners. The cakes were inoculated with powdered cakes from previous batches and spread on a bamboo mat or floor which were already spread with thick layer of husk or straw or sack. After being properly spread, the pieces of ***hamei*** were again covered with thick layers of husk or straw or sack. Fermentation was completed when a characteristic alcoholic smell was released. Then, pieces of ***hamei*** were spread in the open air for a few times and they were ready to be used for preparation of local liquor (**Yu** or **Zouziu**).

3.16.2. Traditional method of 'Yu' (Zouziu) preparation

For wine manufacturing, **Chamang** rice (kind of rice which contains membrane) was cooked and spread on a bamboo mat. ***Hamei*** was crushed and sprinkled over the rice and then thoroughly mixed. The mash was then placed on a large earthenware pot or 'kourphus' (large aluminum vessel) and covered with a cloth or sack and left for 4 to 5 days. Water is added when the mash developed a characteristic alcoholic flavour.

The final mixture was poured into a large aluminum vessel specially used for wine distillation. It was made in such a way that there was a hole on the side of the vessel to insert a pipe. The vessel was heated at a fireplace; alcohol that flowed down through the pipe was collected in a container. At the top of the vessel a cold-water vessel was placed to obstruct the vapour from escaping and act as a condenser. When the water became hot it was replaced by fresh cold water.

3.16.3. Laboratory method of hamei preparation

Fermentation was carried out in the laboratory by a modified form of the traditional method. *Albizia* bark and rice were crushed into powder by mechanical grinder. The powder was packed in a polythene and gamma- sterilised by using cobalt 60 gamma source in a package irradiator (AECL, Canada) (dose-rate 32 Gy/m). The radiation dose of 25 KGy was delivered for complete sterilization.

Isolated mould and yeast strains were grown separately in YPD (yeast- extract, peptone and Dextrose) broth at 26± 2°C and kept overnight in the rotary shaker (150 rpm). One ml of culture was drawn in a sterile eppendorf tube and centrifuged in mini centrifuge (10,000 rpm for 5 min), and resuspended the pellet in 1 ml sterile distilled water.

12.5g of gamma-irradiated rice powder, 0.25g of sterile bark powder, 11.5 ml of sterile distilled water and 1 ml culture of isolates were mixed thoroughly and made into thick paste. The paste was then made into cake and put in the sterile petriplates and incubated at 26 ± 2°C for 5 days. The fermented cake was used as an inoculum in wine manufacturing.

For further experiment, *M. indicus* was maintained on slants/plates of Yeast extract and Peptone Dextrose Agar (YPDA) pH 5.7 at 4°C. Before each experiment, the organisms were transferred by a loop to a freshly prepared YPDA broth of 100 ml (Erlenmeyer flask) and grown overnight in a rotary shaker (150 rpm). One ml of the broth was pipetted, washed with sterile water, centrifuged, and the pellet was resuspended and inoculated in 12.5 gm and 0.25 gm of gamma-irradiated rice and bark powder. *A.kalkora* bark was used in all the experiments.

3.16.4 Laboratory method for rice fermentation

Rice was dehusked by a mixi grinder and the husks were separated by an air blower. Rice (50g) was cooked in 5 volume of water, cooled and mixed with a single cake. The mixture was transfered into a sterile volumetric flask plugged with cotton, wrapped with aluminium foil and incubated for five days at ambient temperature (26±2°C). After incubation the fermented mash was transferred into a bottle and centrifuged in a Supra Refrigerated Centrifuged (Zouan, France) at 6000 rpm for 20 minutes. The supernatant was kept for further analysis.

3.17. Analysis of fermented end product

The supernatant obtained by centrifugation was double distilled at 78°C. Then 60 ml of distillate was analysed by GCMS and quantified by ADH method.

3.17.1. Gas Chromatography and Mass Spectroscopic (GCMS) analysis of distillate was done by the method of Kang et al. (2003)

Headspace volatiles of the distillate were subjected to GC/MS analysis using a Shimadzu, Qp.550A gas chromatography (mass spectrometry instrument). The instrument was equipped with a GC-17A gas chromatography and provided with DB-1 (Dimethyl polysiloxane, J&W Scientific capillary column length: 30mm, internal diameter: 0.25mm and film: 0.25μm). The operating conditions were column-temperature programmed from 60°C to 200°C at the rate of 4°C/min, held at initial temperature and at 200°C for

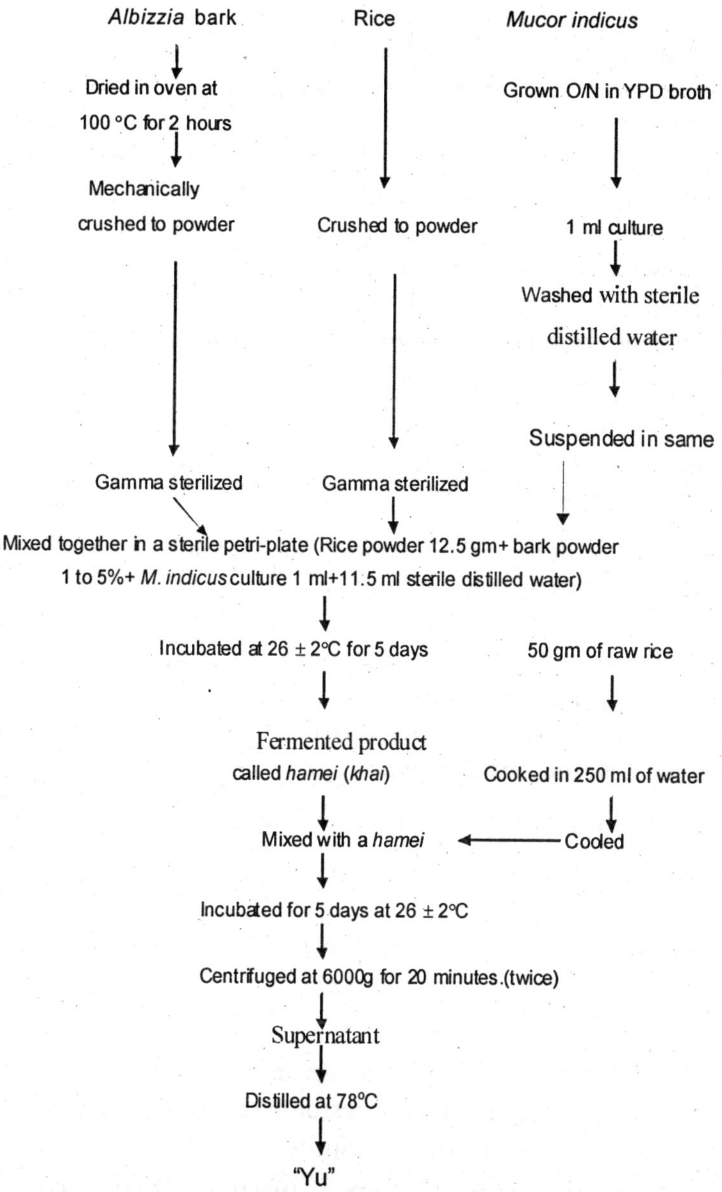

Fig. 4: Process flow chart for the laboratory production of "Yu"

five minutes and further to 280°C at the rate of 10°C /minute held at final temperature for 20 minutes. Injector and interface temperatures were maintained at 210°C and 230°C respectively. Helium was used as carrier gas. Ionization voltage was 70 eV. Electron multiplier voltage was 1 KV. Compounds of interest were identified by comparing their mass fragmentation pattern with that of standard spectra available in the spectral library (flavour and fragrance and Wiley/NIST libraries) of the instrument as well as by comparing the retention times of standards injected separately under the same conditions.

3.17.2 Quantification of ethanol in distilled product (was done by enzymatic assay, Burnt and Gutmann (1974) method)

The distillate, mainly containing ethanol was analysed by GCMS analysis. The actual ethanol content was quantified by ADH (alcohol dehydrogenase) assay. For ADH assay the distillate was diluted to 1000 times (10 μl to 10 ml) using sterile distilled water and from this diluted solution, 10 μl was analysed in 1 ml reaction mixture. Thus, effective dilution in final reaction mixture was 1:10.

Formula for NADH+H$^+$ estimation

According to Beer Lamberts' Law,

NADH + H$^+$ conc. (mM) = absorbance/molar extinction coefficient x path length (cm)

= A/6.22 x1

= A/6.22.

NADH + H$^+$ formation was estimated by measuring the absorbance at 340nm. In a spectrophotometer (UV visible pi Unicam U.K.) Alcohol conc. is equimolar to NADH + H$^+$ conc.

$$RCH_2OH + NAD^+ \xrightarrow{NAD} RCHO + NADH + H^+$$

3.18. Studies of Factors Affecting Ethanol Yield

3.18.1. Determination of ethanol yield from albizia variety, water quantity and mode of inoculation and water quantity in rice cooking

50 gm of rice was cooked in 5 vol. and 10 vol. of water. Rice cooked in 5 volume of water was separately inoculated by ***hamei*** of *A.myriophylla*, ***hamei*** of *A.kalkora*, 1 ml mould (*M.indicus*) suspension, 1 ml mould suspension + *A.myriophylla* bark powder and 1 ml mould suspension + *A.kalkora* bark powder. Similarly, rice cooked in 1:10 volume of water was inoculated in the same manner as mention above. Ethanol was extracted by double distillation at 78°C and estimated by Enzymatic assay.

3.18.2 Determination of optimum bark powder in hamei

Hamei was prepared in various concentration of bark powder from 0 percentage to 5 percentage in the same procedure as described in 3.16.3. ***Hamei*** was inoculated in cooked rice and incubated for 5 days. Ethanol was extracted by double distillation at 78°C and estimated by enzymatic assay.

3.18.3 Determination of optimum incubation period for Hamei fermentation

Hamei was prepared as mentioned in 3.16.3 by KD-2-6-3 rice and *A. kalkora* bark powder, and was harvested from day 1 to 5 and inoculated consecutively into cooked rice (RCM-9 variety) and ethanol was extracted.

Again, 5-day incubated ***hamei*** was inoculated separately in cooked rice as mentioned in 3.16.4. Periodic extractions of ethanol were done from day 1 to 5. Ethanol was extracted as mentioned in 3.17 and estimated by enzymatic assay.

3.18.4 Determination of effect of rice quality and quantity of Hamei inoculum

Hamei was prepared from KD-2-6-3 rice and **Kolam** rice as mentioned in 3.16.3 and inoculated in cooked rice (RCM-

9 and Kolam). Six gm (half) ***Hamei*** cakes were also inoculated in cooked rice and estimation of ethanol from the final product of all the fermentations were done by enzymatic assay.

3.18.5. Determination of the suitable rice variety for Hamei and fermentation

Hamei was prepared from 16 rice cultivars in the same procedure as mentioned in 3.16.3 using pure mould (*M.indicus*) culture and inoculated in 50 gm cooked rice (**Kolam** variety) separately. Yield was determined.

Again ***hamei*** prepared from KD-2-6-3 variety was inoculated separately in cooked rice of 16 different varieties and determined the yield.

Ethanols were extracted by double distillation at 78°C and estimated by enzymatic assay.

3.18.6 Determination of ethanol yield from hamei prepared by phenolic acids substituting Albizia bark powder

This particular ***Hamei*** was prepared by adding a mixture of 2% caffeic acid, 2% ferulic acid, 2% vanillic acid and 2% salicylic acid, and four individual ***hameis*** of 2% caffeic acid, 2% ferulic acid, 2% vanillic acid and 2% salicylic acid. Cakes were inoculated in the cooked rice and fermented. The ethanol yield was compared with the ***hamei*** made of *Albizia* bark powder. Estimation was done by enzymatic assay.

3.19. Morphological and physiological feature of *M.indicus*

M.indicus, which was maintained in YPDA plates/slants at 4°C were reinoculated in YPD broth at pH 5.7 and grown overnight in a rotary shaker (150 rpm). One ml of it was drawn, washed, centrifuged and the pellet resuspended in sterile saline (.85%). It was serially diluted and plated on YPDA plates (pH 5.7) and incubated at ambient temperature

(26±2°C). Growth and other morphological character of *M.indicus* were observed.

3.19.1. Growth kinetics of Mucor indicus in Hamei (cake) and fermentation broth

Hamei was incubated at ambient temperature for 5 days. On each day, 1 gm of cake was cut using a sterile scissors and suspended in 5 volume of sterile water. One ml of aliquot from the suspension was serially diluted using sterile saline (0.85%) and plated on YPDA plates. The plates were incubated at 26 ± 2°C for 48 hours. The number of viable colony forming unit was counted at the end of incubation and CFU/g was calculated taking dilution factor into consideration. Similarly, mould counting was also performed from fermentation broth from day 1 to 5 of incubation by withdrawing 1 ml of the sample, serially diluting with saline (0,85%) and plating on YPDA plates.

3.19.2. Temperature tolerance of Mucor indicus

The organism from actively growing culture plate was streaked on YPDA plate in different replicates. The plates were incubated at various temperatures ranging from 4°C to 60°C for 48 hours and, at the end of incubation period, growths were observed

3.19.3. pH Tolerance of Mucor indicus

The organism from actively growing cultures plates was streaked on YPDA plate in different pH ranging from 5 to10. The plates were incubated at 26 ± 2°C for 48 hours. At the end of incubation, growth was checked. For pH less than 5, assay was performed in YPD broth, because agar was not able to solidify at pH 4 and below.

3.19.4. Studies of ethanol tolerance of Mucor indicus

The mould was grown overnight in 10 ml of YPD (pH 5.7) broth at ambient temperature in a rotary shaker (150 rpm). The culture was centrifuged for 1 minute and the pellet was washed with equal volume of 0.85% saline. The pellet was suspended in the same solution. The cell suspension was mixed with different concentrations of ethanol (Hayman, UK) varying from 5 to 50% and incubated for 2 hours or more. The treated cell suspension was serially diluted using saline, plated on LB agar plate and incubated at 26±2°C for 48 hours and the viable colony forming unit number was counted.

COLOUR PLATES 1-15

Plate 1. Mpangban

Plate 2. Napdai

Plate 3. Banglai

Plate 4. De se

Plate 5. KD 2-6-3

Plate 6. RCM 5

Plate 7. RCM 9

Plate 8. RCM 10

Plate 9. *Albizia kalkora* plant

Plate 10. *Albizia kalkora* plant

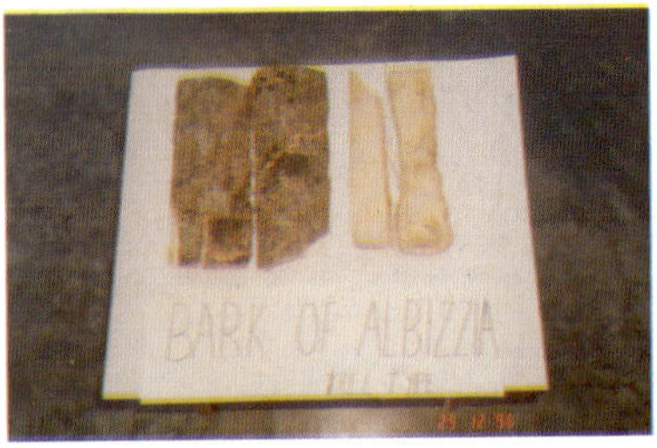

Plate 11. *Albizia kalkora* bark

Plate 12. *Albizia kalkora* bark

Plate 13. *Albizia myriophylla* plant

Plate 14. *Albizia myriophylla* plant

Plate 15. *Albizia myriophylla* bark

4

Results

The result of the role of *Albizia* bark in the fermentation and biochemical analysis of the substrate in different parameters are presented below:

Two types of *Albizia spp.* commonly used by brewers in preparing ***hamei*** were collected and identified as *Albizia myriophylla* Benth which is a woody climber and *A.Kalkora* Pain an erect tree. They grow in the temperate forest of Manipur. The fibrous bark is sweetish and astringent in taste.

Different rice cultivars: RCM-5, RCM-9 and RCM-10 were collected from ICAR, Imphal. KD 2-6-3, IR-64, CAUR-3 and **Tamphaphou** were collected from CAU Imphal. Brewers in Imphal valley commonly use these varieties of rice. **Mpangban, Tunanapnang, Napdai, Banglai, Ritiang, Nappuak, Sangbu, De se** and **Shazei** which are commonly cultivated and used for brewing were collected from Tamenglong District, Manipur.

4.1.1 Moisture Content and Ash Content of Rice Grain

4.1.2 Moisture percent (%)

Moisture content of 16 rice cultivars is depicted in Table 1 and graphically presented in Fig-5. A critical examination of the data reveals that there is a significant variation of moisture content in the grain ranging from 7.4 % to 11.59 %. The maximum content is recorded as 11.59 %, 11.32 % and 11 % in **Mpangban** (T_1), **Tunanapnang** (T_2) and **Napdai** (T_3) respectively. Minimum content is recorded as 7.4 %, 7.45 %, 7,83 % in **Ritiang** (T_5), **De se** (T_9) and **Sangbu** (T_8) respectively.

4.1.3 Ash content (%)

Data for ash content of rice grain is presented in Table 1 and graphically depicted in Fig-5. The examination of the results reveals that there are significant variations of ash content among the grains. From the observation, it is found that variety KD-2-6-3 (T_{10}) has the highest ash content at 1.96 %, followed by RCM-10 (T_{16}) at 1.93 %, IR-64 (T_{13}) 1.90 and **Tunanapnang** (T_2) at 1.88%. The minimum ash content is recorded as 0.96 % in **De se** (T_9) followed by 1.16 % in **Sangbu** (T_8).

4.2. Starch, Amylose and Mylopectin Contents of Rice Grain

4.2.1. Starch content (mg/g)

Table II represents the starch content of 16 rice grains and it is graphically depicted in Fig 6. Starch content of rice grain differs significantly from each other. A critical examination of the results reveals that **Mpangban** (T_1) got the maximum starch content at 88.48 followed by **Tunanapnang** (T_2) 87.95, **Napdai** (T_3) 79.78 and KD-2-6-3 (T_{10}) 78.43. Little variations are observed between RCM-9 (T_{15}) 76.18 and RCM-5 (T_{14}) 76.16. Minimum starch content is observed in IR-64 (T_{13}) as 69.89.

4.2.2. Amylose content (mg/g)

A study of Table II in Fig 6 reveals that there are significant differences in amylose content among different rice cultivars. Maximum amylose content is recorded as 26.87 in IR-64 (T_{13}), 26.71 in **Banglai** (T_4), 26.59 in **De se** (T_9), 26.18 in **Ritiang** (T_5) and 26.08 in **Sangbu** (T_8). Minimum amylose content is observed as 10.11 in **Mpangban** (T_1) followed by 14.28 in **Tunanapnang** (T_2) 15.73 in KD-2-6-3 (T_{10}) and 19.75 in **Napdai** (T_3).

Table I
Percentage of Moisture and Ash content in rice grains

Treatment	*Moisture % in fresh wt. basis*	*Ash % in dry wt. basis*
T_1 MPANGBAN	11.59 + 0.09	1.77 + 0.06
T_2 TUNANAPNANG	11.32 + 0.03	1.88 + 0.06
T_3 NAPDAI	11.0 + 0.55	1.70 + 0.1
T_4 BANGLAI	8.15 + 0.07	1.2 + 0.1
T_5 RITIANG	7.4 + 0.06	1.5 + 0.4
T_6 NAPPUAK	8.25 + 0.09	1.86 + 0.15
T_7 SHAZEI	10.16 + 0.09	1.56 + 0.76
T_8 SANGBU	7.83 + 0.49	1.16 + 0.06
T_9 DE SE	7.45 + 0.54	0.96 + 0.25
T_{10} $KD_{2\text{-}6\text{-}3}$	10.43 + 0.06	1.96 + 0.06
T_{11} TAMPHAPHOU	9.85 + 0.14	1.46 + 0.31
T_{12} CAU R_3	9.05 + 0.06	1.53 + 0.32
T_{13} IR_{64}	8.03 + 0.05	1.90 + 0.1
T_{14} RCM_5	9.19 + 0.24	1.83 + 0.81
T_{15} RCM_9	8.81 + 0.44	1.8 + 0.00
T_{16} RCM_{10}	8.92 + 0.10	1.93 + 0.06
CD at 5%	SE = 0.22 CD = 0.46	SE = 0.26 CD= 0.55

Table II
Starch, Amylose and Amylopectin (mg/ gm) content in rice grain

Treatment	*Starch*	*Amylose*	*Amylopectin (mg/g of total starch extracted*
T_1 MPANGBAN	88.48 ± 0.55	10.11 ± 0.06	89.89 ± 0.06
T_2 TUNANAPNANG	87.95 ± 0.42	14.28 ± 0.06	85.71 ± 0.06
T_3 NAPDAI	79.78 ± 0.39	19.75 ± 0.06	80.25 ± 0.06
T_4 BANGLAI	71.85 ± 0.53	26.71 ± 0.11	73,29 ± 0.11
T_5 RITIANG	70.29 ± 0.12	26.18 ± 0.09	73.82 ± 0.08
T_6 NAPPUAK	71.56 ± 0.89	23.09 ± 0.08	76.91 ± 0.08
T_7 SHAZEI	75.33 ± 0.13	22.4 ± 0.11	77.6 ± 0.11
T_8 SANGBU	70.26 ± 0.23	26.08 ± 0.15	73.92 ± 0.14
T_9 DE SE	74.62 ± 0.13	26.59 ± 0.28	73.41 ± 0.27
T_{10} $KD_{2\text{-}6\text{-}3}$	78.43 ± 0.27	15.73 ± 0.07	84.27 ± 0.07
T_{11} TAMPHAPHOU	73.00 ± 1.04	23.14 ± 0.09	76.86 ± 0.09
T_{12} CAU R_3	72.32 ± 0.06	23.18 ± 0.14	76.82 ± 0.14
T_{13} IR_{64}	69.89 ± 0.06	26.87 ± 0.12	73.13 ± 0.12
T_{14} RCM_5	76.16 ± 0.76	22.71 ± 0.07	77.29 ± 0.07
T_{15} RCM_9	76.18 ± 0.37	22.89 ± 0.13	77.11 ± 0.13
T_{16} RCM_{10}	76.40 ± 1.26	22.96 ± 0.23	77.03 ± 0.22
CD at 5%	SE=0.51 CD=1.09	SE=0.07 CD =0.14	SE=0.10 CD=0.21

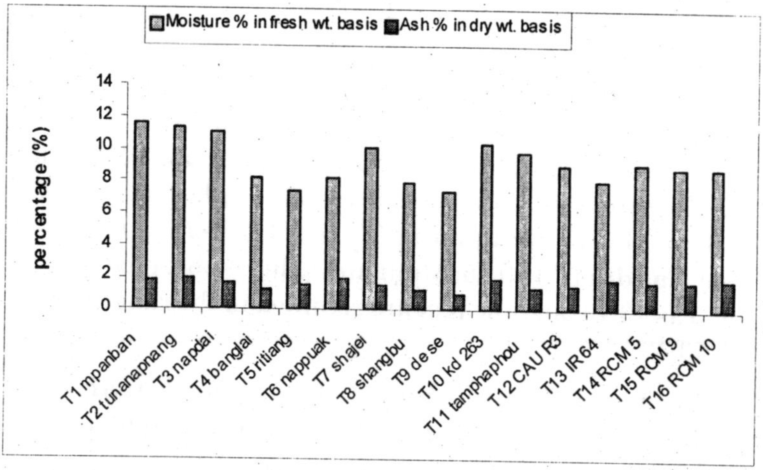

Fig. 5: Percentage of moisture and ash in rice grain

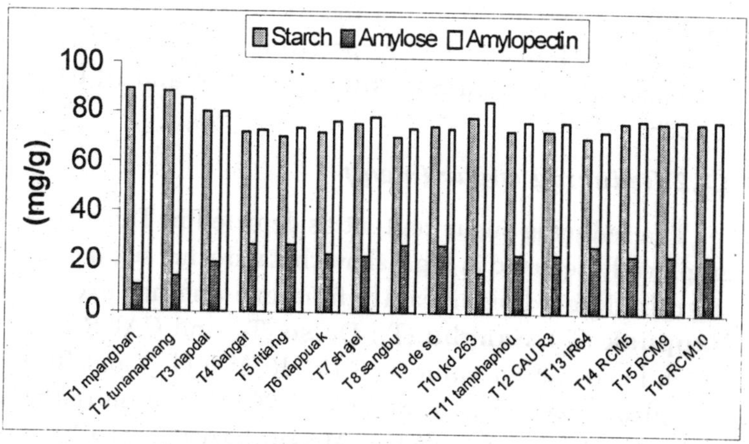

Fig. 6: Level of starch, amylose and amylopectin in rice grain

4.2.1. Amylopectin content (mg/g)

Amylopectin content is depicted in Table II and graphically presented in Fig 6. Critical analysis of the result reveals that there are significant variations in the amylopectin content among 16 rice cultivars. Maximum amylopectin content is recorded in **Mpangban** (T_1) as 89.89 followed by **Tunanapnang** (T_2) 85.71, KD-2-6-3 (T_{10}) 84.27 and **Napdai** (T_3) 80.25. Minimum value is observed in IR-64 (T_{13}) as 73.13, and the rest are recorded in the range of 89.89 to 73.13.

4.3.0. Total Soluble Sugar, Reducing Sugar, Non-reducing Sugar and Total Free Amino Acid in Rice Grain

4.3.1 Total soluble sugar (mg/g)

Results presented in table III and graphically depicted in Fig 7 on total soluble sugar content of rice grains show a remarkable difference among the grains. Analysis of the results reveals the highest content as 1.17 in RCM-9 (T_{14}). **Mpangban** (T_1), RCM-5 (T_{14}) and RCM-10 (T_{16}) are found to contain 1.16, 1.13 and 1.12 respectively. However minimum content is observed in **Sangbu** (T_8) as 0.66 and IR-64 (T_{13}) as 0.68.

4.3.2 Reducing sugar (mg/g)

The reducing sugar content as depicted in Table III and graphically depicted in Fig.7 show that reducing sugar content of grain differed significantly from one another except for **Nappuak** (T_6), **Sangbu** (T_8) **De se** (T_9) and CAUR-3 (T_{12}), which have similar value as 0.19; RCM-5 (T_{14}) and RCM-9 (T_{15}) also have similar content as 0.28. Maximum reducing sugar content is observed in **Mpangban** (T_1) at 0.52 and next in **Tunanapnang** (T_2) as 0.47 and 0.42 in KD-2-6-3 (T_{10}); IR-64 (T_{13}) has the minimum value at 0.14.

Table III

Total soluble sugar, Reducing sugar, Non-reducing sugar and Total free amino acid (mg/g) in rice grain

Treatment	*% in dry weight*			*Total free amino acid*
	Total soluble sugar	*Reducing sugar*	*Non-reducing sugar*	
T_1 MPANGBAN	1.16 + 0.06	0.52 + 0.02	0.64 + 0.07	1.56 + 0.005
T_2 TUNANAPNANG	1.04 + 0.03	0.47 + 0.02	0.57 + 0.04	1.40 + 0.14
T_3 NAPDAI	1.04 + 0.05	0.37 + 0.03	0.68 + 0.07	0.55 + 0.00
T_4 BANGLAI	0.99 + 0.02	0.38 + 0.03	0.61 + 0.03	0.06 + 0.05
T_5 RITIANG	0.83 + 0.04	0.21 + 0.02	0.62 + 0.05	0.83 + 0.02
T_6 NAPPUAK	0.81 + 0.06	0.19 + 0.03	0.62 + 0.04	1.15 + 0.01
T_7 SHAZEI	0.94 + 0.04	0.30 + 0.01	0.64 + 0.05	0.99 + 0.005
T_8 SANGBU	0.66 + 0.03	0.19 + 0.02	0.47 + 0.02	1.34 + 0.005
T_9 DE SE	0.86 + 0.02	0.19 + 0.02	0.67 + 0.02	1.56 + 0.01
T_{10} KD2-6-3	1.06 + 0.01	0.42 + 0.04	0.64 + 0.03	0.52 + 0.02
T_{11} TAMPHAPHOU	0. 91 + 0.01	0.22 + 0.03	0.69 + 0.04	0.54 + 0.02
T_{12} CAU R3	0.84 + 0.05	0.19 + 0.02	0.65 + 0.04	0.47 + 0.01
T_{13} IR 64	0.68 + 0.04	0.14 + 0.03	0.54 + 0.05	1.29 + 0.02
T_{14} RCM5	1.13 + 0.04	0.28 + 0.01	0.85 + 0.04	1.00 + 0.02
T_{15} RCM9	1.17 + 0.03	0.28 + 0.02	0.89 + 0.05	0.82 + 0.02
T_{16} RCM10	1.12 + 0.04	0.32 + 0.03	0.80 + 0.07	0.56 + 0.02
CD at 5%	SE = 0.026 CD = 0.055	SE = + 0.018 CD = 0.039	SE = 0.036 CD = 0.077	SE = 0.03 CD = 0.06

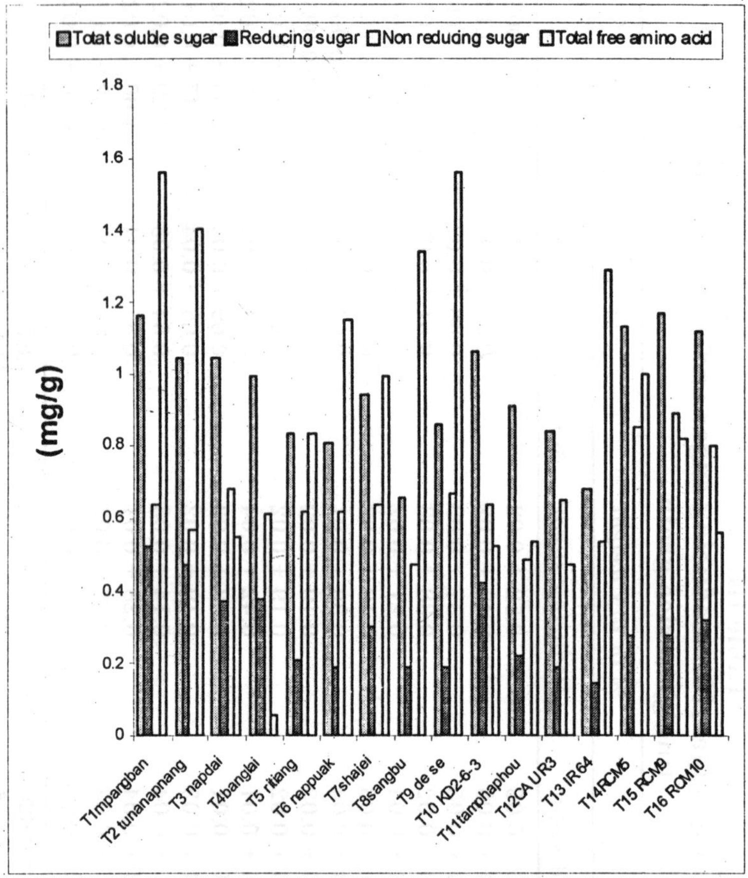

Fig. 7: Level of total soluble sugar, reducing sugar, non-reducing sugar and total free amino acid in rice grain

4.3.1 Non Reducing Sugar (mg/g)

A study of Table III and Fig.7 representing non-reducing sugar content of 16 rice grains reveals that there are significant variations among the grains. However insignificant differences are recorded in **Ritiang** (T_5) at 0.62, **Nappuak** (T_6) at 0.62, **Mpangban** (T_1) at 0.64, **Shazei** (T_7) at 0.64 and KD-2-6-3 (T_{10}) at 0.64. Maximum non-reducing sugars are observed in RCM-9 (T_{14}) at 0.89 and in RCM-5(T_{14}) at 0.85. Minimum content is observed in **Sangbu** (T_8) at 0.47.

4.3.4 Total free amino acid (mg/g)

Total free amino acid content of rice grains is depicted in Table III and graphically depicted in Fig7. Data presented in the table reveal that there are significant differences of total amino acid among the rice grains. Maximum content is observed in **Mpangban** (T_1) 1.56 and **De se** (T_9) 1.56. Minimum content is observed in CAUR-3 (T_{12}) 0.47 and KD-2-6-3 (T_{10}) 0.52.

4.3.4.0. Chemical Composition of *Albizia* Bark

4.3.4.1. Moisture content

Data for moisture content of *Albizia* bark are depicted in Table IV which show remarkable differences in *A.myriophylla* and *A. kalkora*. It is also graphically depicted in Fig 8. Critical examination of the data reveals that there is a significant variation of moisture content. Higher moisture level is observed in *A.kalkora* 39.65 % than in *A. myriophylla* 27.10 %.

4.3.4.2. Ash Content

Results presented in Table IV and graphically depicted in Fig. 8 on ash content of *Albizia myriophylla* and *A. kalkora* barks clearly indicate a significant difference. Higher ash content is observed in *A. kalkora* at 32.08% than in *A.myriophalla* at 29.13 %.

4.3.4.3. Total free Amino acid (mg/g)

Data for free amino acid content of *Albizia* bark are recorded in Table IV and graphically presented in Fig.9. It is found to be significantly different between the two species. From the observation it is revealed that amino acid content is higher in *A. myriophylla* (0.23 mg/g) than *A. kalkora* (0.07 mg/g).

4.3.4.4. Protein mg/g

It is evident from Table IV and Fig.8a that crude protein contents of *A. myriophylla* and *A. kalkora* are significantly different. Higher values are observed in *A. myriophylla* at 0.19 mg/g than in *A. kalkora* at 0.02mg/g.

4.4.1. Total Phenol (mg/g)

The total phenol content of *Albizia* bark is presented in Table IV and graphically depicted in Fig.9. The analyses of the results indicate that there are remarkable difference in the total phenol content of *A. kalkora* and *A. myriophylla* which is found to be 19.02 and 12.47 respectively.

4.4.1.1. Acidic nature of Albizia bark

Table V and Fig.10 depict the acidic nature of the *Albizia* bark (*A. myriophylla and A. kalkora*) and its pH value are found within the range of 5.03 to 4.86. pH value for bark concentration at 0.5%, 1%, 2%, 3%, 5% and 10% is determined and found to be 5.03 and 5.09 at 5% concentration, 4.92 and 4.94 at 1% concentration respectively. At 3% and up to 10% concentration, no significant variable is observed and the value remains constant even if the concentration is increased.

4.4.1.2. Organic acid

Flow chart of organic acid extraction of *Albizia* bark is depicted in Fig.1. The extracted samples are analysed for organic acid by thin layer chromatography (TLC). From the analysis of the results it is confirmed that no organic acid are not detected from the extracted sample.

Table IV
Chemical composition of Albizia bark

Treatment	*Moisture % in fresh wt.*	*Ash % in dry wt.*	*Total free amino acid mg/g*	*Protein mg/g*	*Total phenol mg/g*
A. myriophylla	27.10 + 0.25	29.13 + 0.25	0.23 + 0.007	0.19 + 0.004	12.47 + 0.06
A. kalkora	39.65 + 0.27	32.08 + 0.29	0.07 + 0.006	0.12 + 0.006	19.02 + 0.06
CD at 5%	SE = 0.09 CD = 0.19	SE = 0.061 CD = 0.13	SE = 0.003 CD = 0.008	SE = 0.002 CD = 0.008	SE = 0.022 CD= 0.05

Table V
Acidic nature of *Albizia* bark

	pH of bark in different concentration					
Bark Percentage	0.5	1	2	3	5	10
Treatment						
Albizia myriophylla	5.03	4.92	4.87	4.86	4.86	4.86
Albizia kalkora	5.09	4.94	4.86	4.84	4.84	4.85

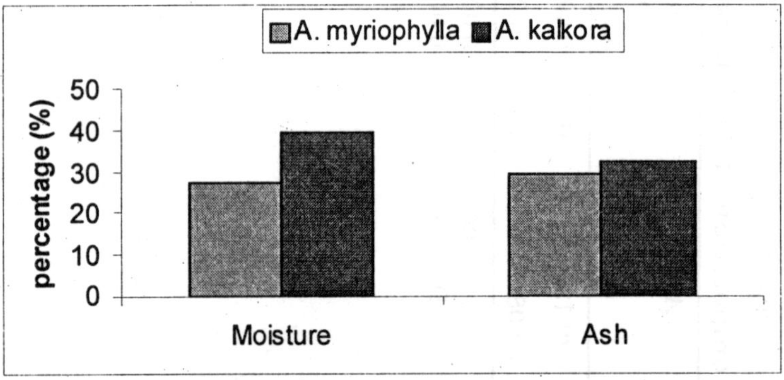

Fig. 8: Moisture and ash content of *A. myriophylla* and *A. kalkora* bark powder

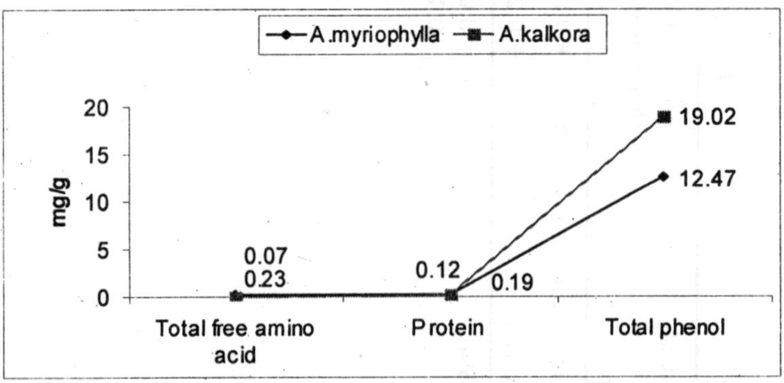

Fig.9 : Total free amino acid, protein and total phenol content in *A. myriophylla* and *A .kalkora* bark powder

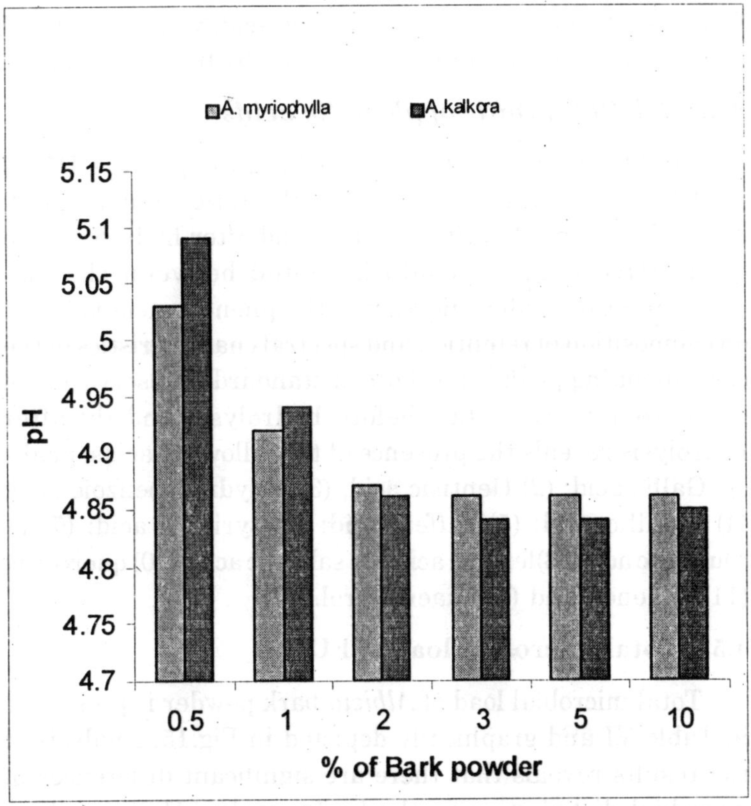

Fig. 10 : Acidic nature of *Albizia* bark

4.4.1.2. Phenolic acid

The flowchart of methanol extraction of phenolic acids and hydrolysis of *Albizia* bark powder is depicted in Fig 2 and 3. The extracts are analysed by thin layer chromatography (TLC). The phenolic acids are identified by colour reaction with spray reagent. Rf value of the sample when compared with the authentic standards sample reveals the presence of phenolic acid in the bark. Based on TLC of bark powder it demonstrates the presence of caffeic acid, ferulic acid, syringic acid, p-coumaric acid, vanillic acid and salicylic acid in both unhydrolysed and hydrolysed samples.

4.4.1.3. HPLC profile of phenolic acids

Fig.11 and Fig.12, Fig.13 and Fig.14 depict the HPLC profile of methanol extract of phenolic acid from *A. myriophylla* and *A. kalkora* before and after hydrolysis. No qualitative changes could be noted between the two chromatograms. Identification of the phenolic acid is based on composition of retention and spectral characteristics of the corresponding peaks and those of standards. Observation of the chromatograms (a) before hydrolysis and (b) after hydrolysis reveals the presence of the following acids; peaks (1) Gallic acid; (2) Gentisic acid; (3) p-hydroxybenzoic acid; (4) vanillic acid; (5) caffeic acid; (6) syringic acid; (7) p-coumaric acid; (8)ferullic acid; (9) salicylic acid; (10) quercetin; (11) eugenol; and (12) kaempferol.

4.5.1 Total microbial load (CFU/g)

Total microbial load of *Albizia* bark powder is presented in Table VI and graphically depicted in Fig.15. Analysis of the results reveals that there are significant differences of microbial distribution in the bark powder. Higher colony forming units (cfu) of the microbes are observed in yeast extract dextrose Agar (PDA) petriplates (media) than Luria Bertani (LB) petriplates (media). Total microbial load in *A. kalkora* bark powder is found to be $9X10^3$ cfu/g in LB media and $1.88X10^4$cfu/g in YPDA which is higher than *A. myriophylla* bark powder which are 80 cfu/g in LB media and $1.2\ X10^2$ cfu/g in YPDA media.

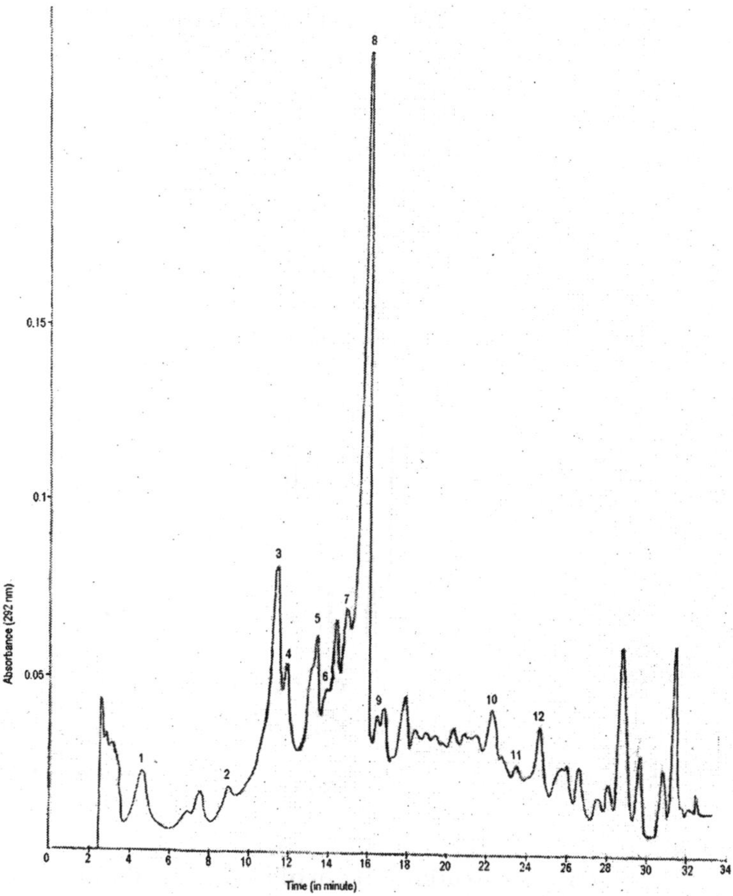

Fig. 11. HPLC Chromatogram of free phenolic acids from *A. myriophylla*

1.	**Gallic acid**	2.	**Gentisic acid**
3.	**p-hydroxybenzoic acid**	4.	**Vanillic acid**
5.	**Caffeic acid**	6.	**Syringic acid**
7.	**P-coumaric acid**	8.	**Ferulic acid**
9.	**Salycylic acid**	10.	**Quercetin**
11.	**Eugenol**	12.	**Kaempferol**

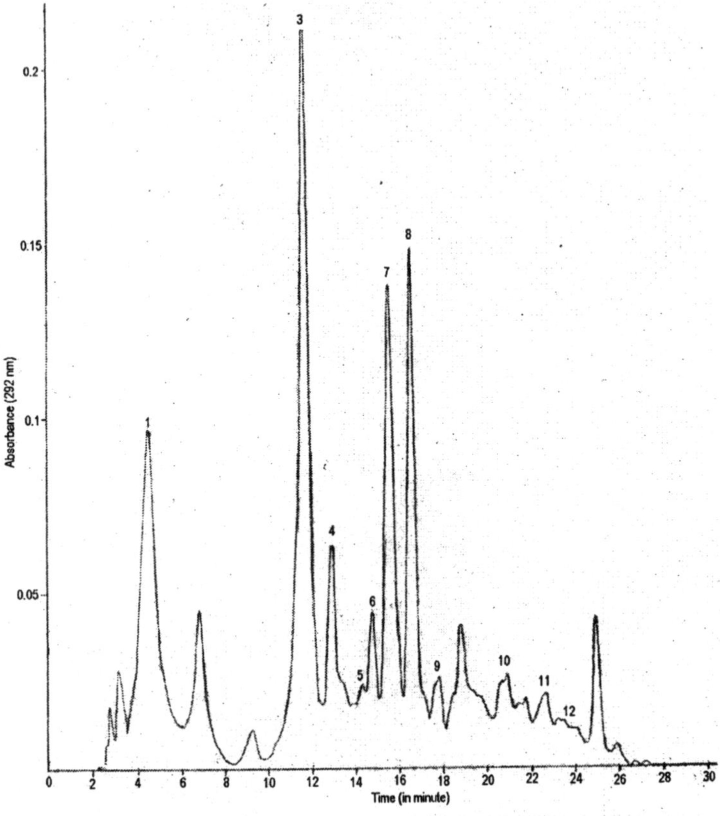

Fig. 12. HPLC Chromatogram of hydrolyzed phenolic acids from *A. myriophylla*

1.	**Gallic acid**	2.	**Gentisic acid**
3.	**Vanillic acid**	4.	**Caffeic acid**
5 .	**Syringic acid**	6.	**P-coumaric acid**
7.	**Ferulic acid**	8.	**Synapic acid**
9.	**Salycylic acid**	10.	**Quercetin**
11.	**Eugenol**	12.	**Kaempferol**

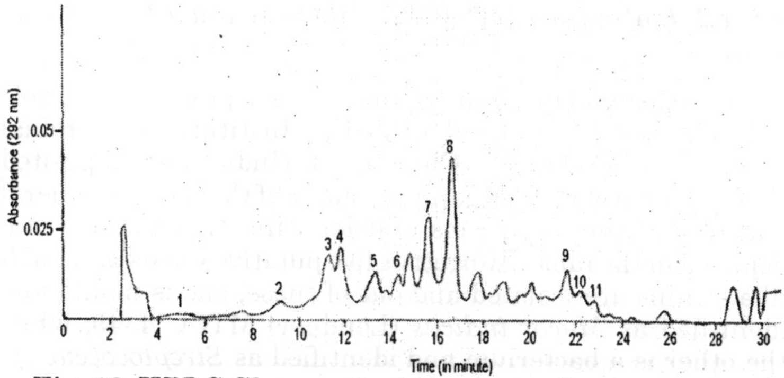

Fig. 13. HPLC Chromatogram of free phenolic acids from *A. kalkora*

1. Gallic acid
2. Gentisic acid
3. p-hydroxybenzoic acid
4. Vanillic acid and caffeic acid
5. Syringic acid
6. P-coumaric acid
7. Ferulic acid
8. Salycylic acid
9. Quercetin acid
10. Eugenol
11. Kaempferol

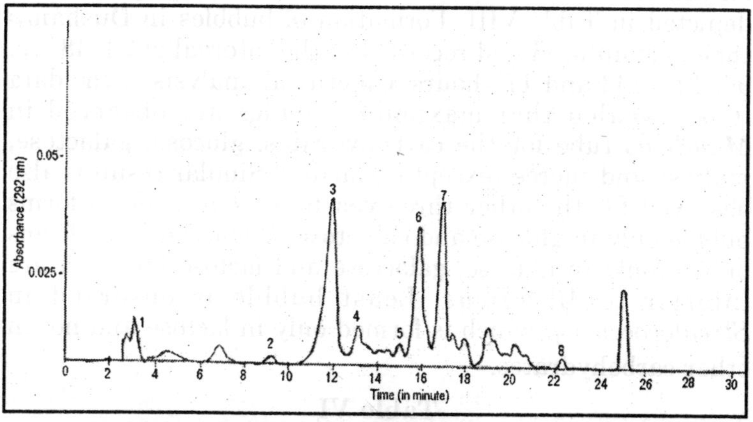

Fig. 14. HPLC Chromatogram of hydrolyzed phenolic acids from *A. kalkora*

1. Gallic acid
2. Gentisic acid
3. Vanillic acid and caffeic acid
4. Syringic acid
5. P-coumaric acid
6. Ferulic acid
7. Salycylic acid
8. Quercetin acid
9. Quercetin acid

4.5.1.1 Isolation and Characterization of microorganisms

Microbial isolates from bark powder are presented in Table VII. The isolates were identified at Institute of Microbial Technology (IMTECH), Chandigarh (India) and deposited there. Source of microorganisms, name of the species, general features of the organisms and identification number are depicted in the table. Altogether five putative yeasts or mould-like strains are isolated and out of these, one is mould and identified as *Mucor indicus* (Lendner) MTCC 4349, while the other is a bacterium and identified as *Streptococcus sp.* MTCC (number not provided). The other three are yeasts, i.e. *Candida parapsilosis* (MTCC 4448), *Rhodotorula mucilaginosa* (MTCC 4449) and *Rhodotorula mucilaginosa* (MTCC 4450).

4.5.1.2 Fermentation of individual carbohydrates by isolated microorganisms

Test for fermentative abilities of isolated microorganisms on individual carbohydrates at different time period are depicted in Table VIII. Formation of bubbles in Dusham's tube is monitored and recorded at the interval of 24, 48, 72, 96, 120, 144 and 170 hours. On critical analysis of the data it is revealed that maximum bubbles are observed in *M.indicus* tube for the carbohydrates, glucose, galactose, maltose and sucrose except for lactose. Similar result is also observed for the other three yeasts. *C. Parapsilosis* forms bubble only in glucose and galactose, *R.mucilaginosa* forms bubble only in glucose, galactose and lactose but both are inferior to *M.indicus*. Least bubble is observed in *Streptococcus sp* which is formed only in lactose and not in other carbohydrates.

Table VI

Total microbial load of *Albizia* bark powder (CFU/g)

Treatment	*Culture medium*	
	Luria Bertani (LB) plate	*Yeast Extract Dextrose Agar (YPDA) plate*
Albizia myriophylla	80	11.2 X 10^4
Albizia kalkora	9 X 10^3	1.88 X 10^4

Table VII
Microbial isolates of *Albizia* bark powder

MicrobesIsolates	*Source*	*Name of the species*	*ID No.*	*General feature*
Mould	*Albizia kalkora*	*Mucor indicus* Lendner	MTCC 4349	Colourless, fast growing & sporulating
Yeast	*Albizia myriophylla*	*Rhodotorula mucilaginosa*	MTCC4449	Pink colour, medium growing
Yeast	*Albizia kalkora*	*Rhodotorula mucilaginosa*	MTCC 4450	Pink colour, medium growing
Yeast	Hamei Cake	*Candida parapsilosis*	MTCC 4448	White colour colony, medium growing
Bacteria	*Albizia kalkora*	*Streptococcus sp.*	Not specify	Colourless, slow growing

Table VIII
Fermentation of individual Carbohydrates by isolated Microorganisms

24hrs

Treatment		*Carbohydrates (2% each)*				
		Glucose	*Galactose*	*Lactose*	*Maltose*	*Sucrose*
Mucor	Turbidity	—	—	—	—	—
Indicus	Gas bubble	+	+	+	+	+
Candida	Turbidity	—	—	—	—	—
Parapsilosis	Gas bubble	—	—	—	—	—
	Turbidity	—	—	—		—
Streptococcus spp	Gas bubble	—	—	—	+	—
Rhadotorula	Turbidity				—	—
Mucilaginosa	Gas bubble	+	+	+	—	—

Table VIII(a)

48hrs

Treatmen		*Carbohydrates (2% each)*				
		Glucose	*Galactose*	*Lactose*	*Maltose*	*Sucrose*
Mucor	Turbidity	ppt	+	+	+	—
Indicus	Gas bubble	+	+	—	+	+
Candida	Turbidity	ppt	ppt	ppt	ppt	ppt
Parapsilosis	Gas bubble	++	+	—	—	—
Streptococcus sp.	Turbidity	+	+	+	+	+
	Gas bubble	—	—	+	—	—
Rhodotorula	Turbidity	+ b	+ b	+ b	+ b	+ b
Mucilaginosa	Gas Bubble	+	+	+	—	—

Table VIII(b)

72hrs

Treatmen		*Carbohydrates (2% each)*				
		Glucose	*Galactose*	*Lactose*	*Maltose*	*Sucrose*
Mucor	Turbidity	+ t	+ t	+ t	+ t	+ t
Indicus	Gas bubble	+ +	+ +	–	+	+
Candida	Turbidity	ppt	ppt	ppt	ppt	ppt
Parapsilosis	Gas bubble	+ +	+	–	–	–
Streptococcus sp.	Turbidity	+ b	+ b	+ b	+ b	+ b
	Gas bubble	–	–	+	–	–
Rhodotorula	Turbidity	+ b	+ b	+ b	+ b	+ b
Mucilaginosa	Gas bubble	+	+	+	–	–

TABLE VIII (c)

96hrs

Treatmen		Carbohydrates (2% each)				
		Glucose	Galactose	Lactose	Maltose	Sucrose
Mucor	Turbidity	+ t	+ t	+ t	+ t	+ t
Indicus	Gas bubble	+ + +	+ + +	–	+ + +	+
Candida	Turbidity	+ b	+ b	+ b	+ b	+ b
Parapsilosis	Gas bubble	+ +	+ +	–	–	–
Streptococcus sp.	Turbidity	+ b	+ b	+ b	+ b	+ b
	Gas bubble	–	–	+	–	–
Rhodotorula	Turbidity	+ b	+ b	+ b	+ b	+ b
Mucilaginosa	Gas bubble	+	+	+ +	–	–

Table VIII(d)

120 hrs

Treatmen		Carbohydrates (2% each)				
		Glucose	*Galactose*	*Lactose*	*Maltose*	*Sucrose*
Mucor	Turbidity	+ t	+ t	+ t	+ t	+ t
Indicus	Gas bubble	+ + +	+ + +	–	+ + +	++
Candida	Turbidity	+ b	+ b	+ b	+ b	+ b
Parapsilosis	Gas bubble	+	+ +	–	–	–
Streptococcus sp.	Turbidity	+ b	+ b	+ b	+ b	+ b
	Gas bubble	–	–	–	–	–
Rhodotorula	Turbidity	+ b	+ b	+ b	+ b	+ b
Mucilaginosa	Gas bubble	+	+ +	+ +	–	–

Table VIII(e)

144hrs

Treatmen		Carbohydrates (2% each)				
		Glucose	*Galactose*	*Lactose*	*Maltose*	*Sucrose*
Mucor	Turbidity	+ t	+ t	+ t	+ t	+ t
Indicus	Gas bubble	+ +	+ + +	–	+ +	+
Candida	Turbidity	+ b	+ b	+ b	+ b	+ b
Parapsilosis	Gas bubble	+	+ +	–	–	–
Streptococcus sp.	Turbidity	+ b	+ b	+ b	+ b	+ b
	Gas bubble	–	–	–	–	–
Rhodotorula	Turbidity	+ b	+ b	+ b	+ b	+ b
Mucilaginosa	Gas bubble	+	+	+ +	–	–

Table VIII(f)

170hrs

Treatmen		*Carbohydrates (2% each)*				
		Glucose	*Galactose*	*Lactose*	*Maltose*	*Sucrose*
Mucor	Turbidity	+ t	+ t	+ t	+ t	+
Indicus	Gas bubble	+ +	+ + +	–	+	+
Candida	Turbidity	+ b	+ b	+ b	+ b	+ b
Parapsilosis	Gas bubble	+	+	–	–	–
Streptococcus sp.	Turbidity	+ b	+ b	+ b	+ b	+ b
	Gas bubble	–	–	–	–	–
Rhodotorula	Turbidity	+ b	+ b	+ b	+ b	+ b
Mucilaginosa	Gas bubble	+	+	+	–	–

–, no observation.
+, represents visible bubble; increasing numbers of (+) indicate great volume of visible bubble.
ppt, precipitation.
+t, precipitation on top.
+b, precipitation at bottom.

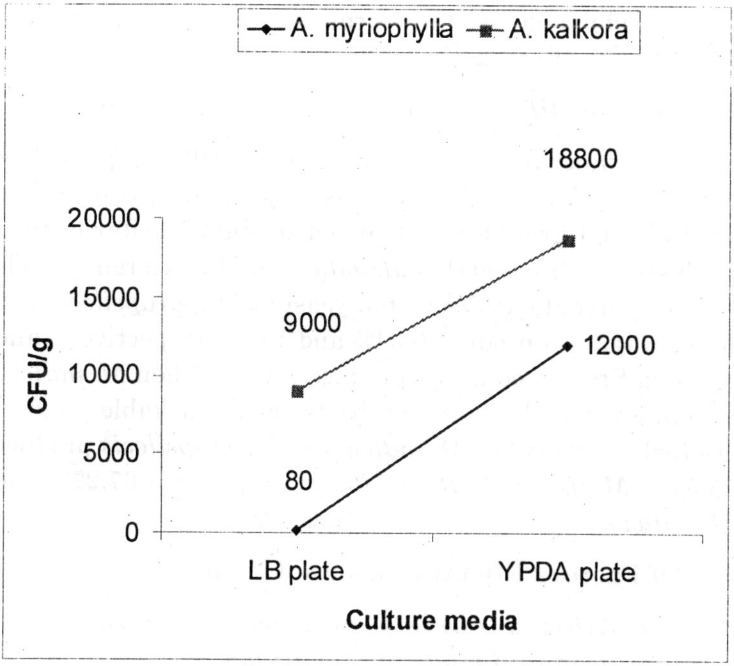

Fig. 15: Total microbial load of *Albizia* bark powder

Another remarkable observation is that bubble formation increases with the increase of time; but after 120 hours, it starts decreasing.

4.5.1.1 Fermentation of rice by isolated microorganisms

Test for rice fermentation abilities of all the isolates are conducted and chromatograms of headspace volatile compound (ethanol) of fermented product analysed by GCMS are shown in Fig.16-26. It is observed that the peak of the distillate obtained from rice fermentation using mould-inoculated (*M. indicus*) ***hamei*** is found to match with the ethanol peak available in the GCMS library (Fig.16,17). Fermentation using yeast and bacteria isolates does not give any resembling peak (Fig. 18,19, 20, 21) but from mould

+yeast and mould +bacteria fermentation, resembling peaks are obtained (Fig. 22,23, 24, 25, 26).

4.5.1.4. Quantification of ethanol in distill product

Data of ethanol percentage of the distillates quantified by ADH (alchohol dehydrogenase) assay are presented in Table X and graphically depicted in Fig.27. After critical analysis it is observed that *M. indicus* produces a remarkable ethanol percentage 65.43% but yeast viz. *C.parapsilosis* and *R.mucilaginosa* produce 0.34% and 1.60% respectively, and bacteria *Streptococcus sp.* produce 0.14%. When *M. indicus* is combined with yeast and bacteria, remarkable yield of ethanol is observed: *M. indicus* + *C. parapsilosis* produce 65.51%, *M. indicus* + *R. mucilaginosa* produce 67.22% and *M. indicus* + *Streptococcus* produce 62.49%.

4.5.2.0 Factors Effecting Ethanol Yield

4.5.2.1 Albizia variety and water quantity in fermentation and mode of inoculums

Results of ethanol yield effected by *Albizia* variety in ***hamei*** mode of inoculums and water volume in fermentation are depicted in Table X and also graphically depicted in

Table IX

Ethanol production from Hamei of different isolated microorganisms (mould, yeast and bacteria)

Treatment	*Hamei type*	*Ethanol %*
T_1	Hamei of *M. indicus*	65.43 ± 0.23
T_2	Hamei of *M. indicus*+ *C. parapsilosis*	65.51 ± 0.24
T_3	Hamei of *M. indicus* + *Septococcs sp.*	62.49 ± 0.35
T_4	Hamei of *M. indicus* + *R.mucilaginosa*	67.22 ± 0.09
T_5	Hamei of *Streptococcus sp.*	00.14 ± 0.11
T_6	Hamei of *C.parapsilosis*	00.34 ± 0.30
T_7	Hamei of *R.mucialginosa*	01.60 ± 0.48
CD at 5%		SE = 0.25
		CD = 0.54

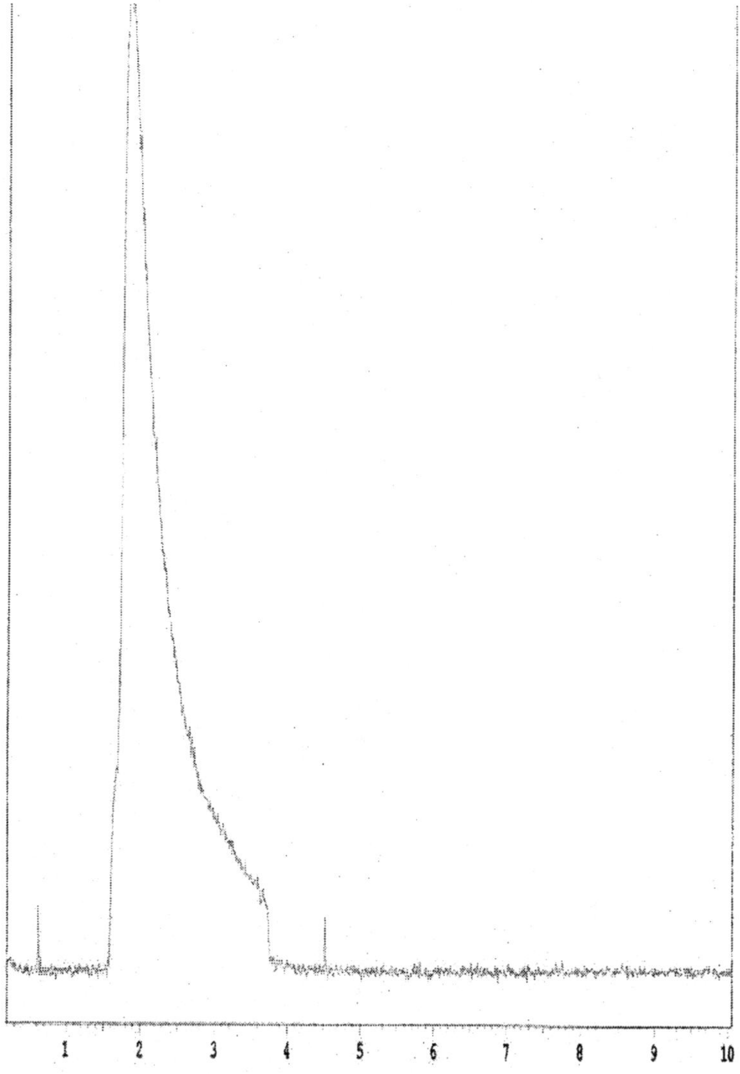

Fig. 16 Chromatograms of headspace volatile compounds (ethanol) of rice fermented product by *M. indicus*

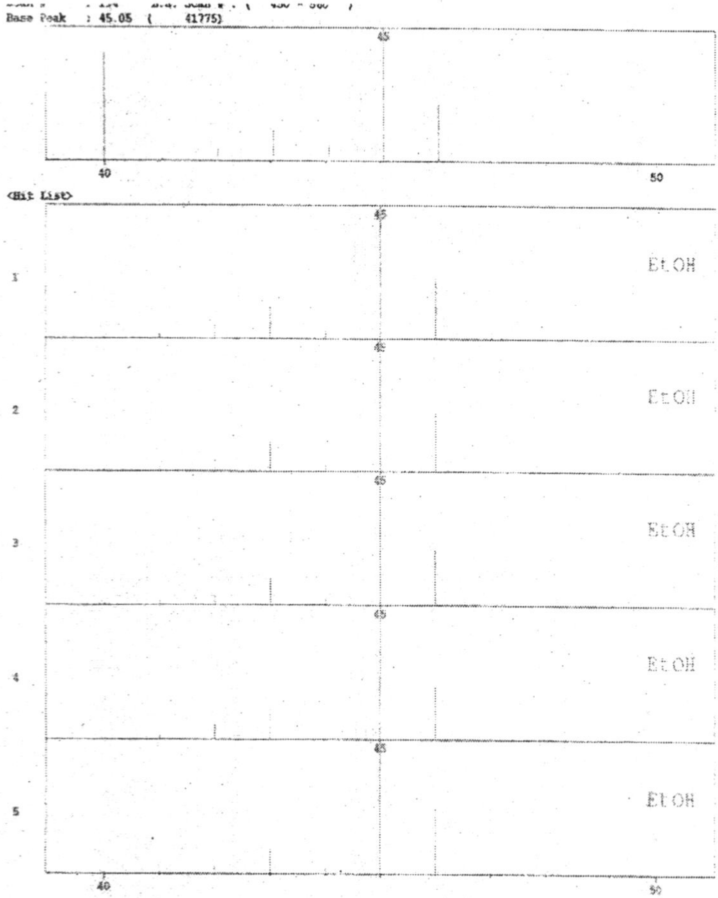

No	SI	Mol.Wgt.	Mol.Form./Compound Name	CAS No.	Entry	LIB#
1	96	46	C_2H_6O	64-17-5	249	2
			Ethanol (CAS) Ethyl alcohol $$ EtOH $$ Tecsol $$ Jaysol $$ Alcohol $$ Algrain $$ Anhy			
2	90	46	C_2H_6O	64-17-5	247	2
			Ethanol (CAS) Ethyl alcohol $$ EtOH $$ Tecsol $$ Jaysol $$ Alcohol $$ Algrain $$ Anhy			
3	95	46	C_2H_6O	64-17-5	241	2
			Ethanol (CAS) Ethyl alcohol $$ EtOH $$ Tecsol $$ Jaysol $$ Alcohol $$ Algrain $$ Anhy			
4	89	46	C_2H_6O	64-17-5	248	2
			Ethanol (CAS) Ethyl alcohol $$ EtOH $$ Tecsol $$ Jaysol $$ Alcohol $$ Algrain $$ Anhy			
5	69	46	C_2H_6O	64-17-5	238	2
			Ethanol (CAS) Ethyl alcohol $$ EtOH $$ Tecsol $$ Jaysol $$ Alcohol $$ Algrain $$ Anhy			

Library Name
(1) SZTERP.LIB (2) WILEY229.LIB

Fig. 17 GCMS Profiel of Volatile compound (ethanol) of rice fermented product by *M. indicus*

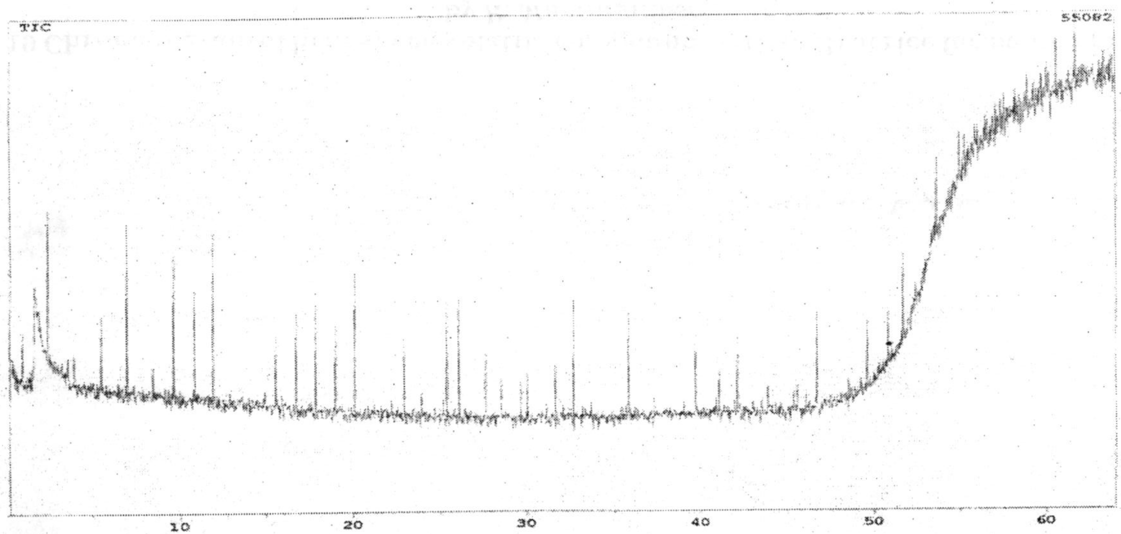

Fig. 18 Chromatogram of headspace volatile compound (ethanol) of rice fermented product by yeast (*C. Parapsilosis*)

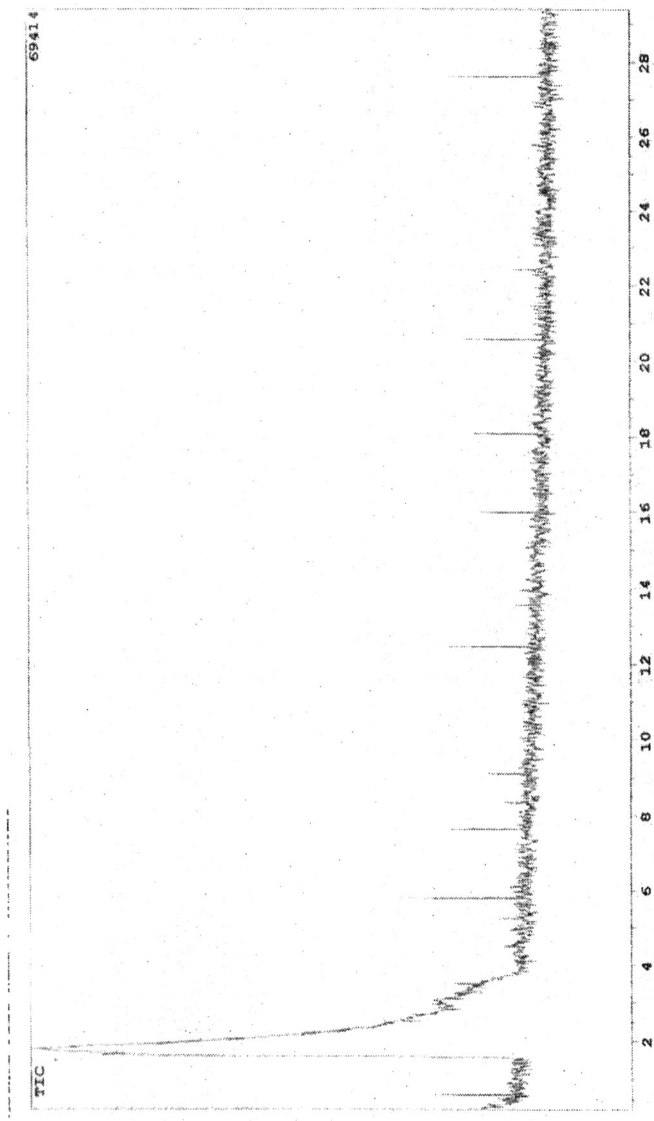

Fig. 19 Chromatogram of headspace volatile compounds (ethanol) of rice fermented product by *R. Mucilaginosa*

Fig. 20 Chromatogram of headspace volatile compounds (ethanol) of rice fermented product by *Streptococcus SP*.

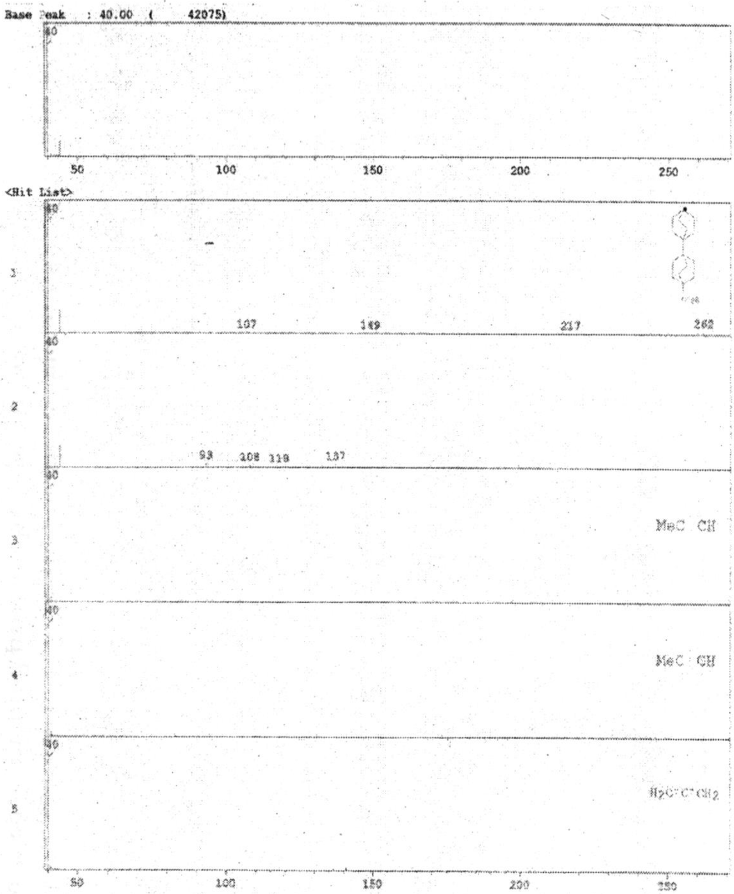

No	SI	Mol.Wgt.	Mol.Form./Compound Name	CAS No.	Entry	LIB#
1	96	262	$C_{17}H_{26}O_2$ 1,1'-Bibicyclo(2.2.2)octyl-4-carboxylic acid $$ [1,1'-Bibicyclo[2.2.2]octane]-4-carbo	74467-50-8	118579	2
2	95	137	$C_8H_{11}NO$ 7-hydroxy-5,6,7,8-tetrahydroindolizinine $$ 7-Indolizinol, 5,6,7,8-tetrahydro- $$	106681-29-1	19929	2
3	94	40	C_3H_4 1-Propyne (CAS) Propyne $$ Propine $$ Allylene $$ Methylacetylene $$	74-99-7	115	2
4	94	40	C_3H_4 1-Propyne (CAS) Propyne $$ Propine $$ Allylene $$ Methylacetylene $$	74-99-7	114	2
5	94	40	C_3H_4 1,2-Propadiene (CAS) Allene $$ Propadiene $$ sym-Allylene $$ Propa-1,2-diene $$ Dimet	463-49-0	118	2

Library Name
(1) S2TERP.LIB (2) WILEY229.LIB

Fig. 21 GCMS profiel of volatile compound of rice fermented product by C. Parapsilosis, R. Mucilaginosa and *Streptococcus SP.*

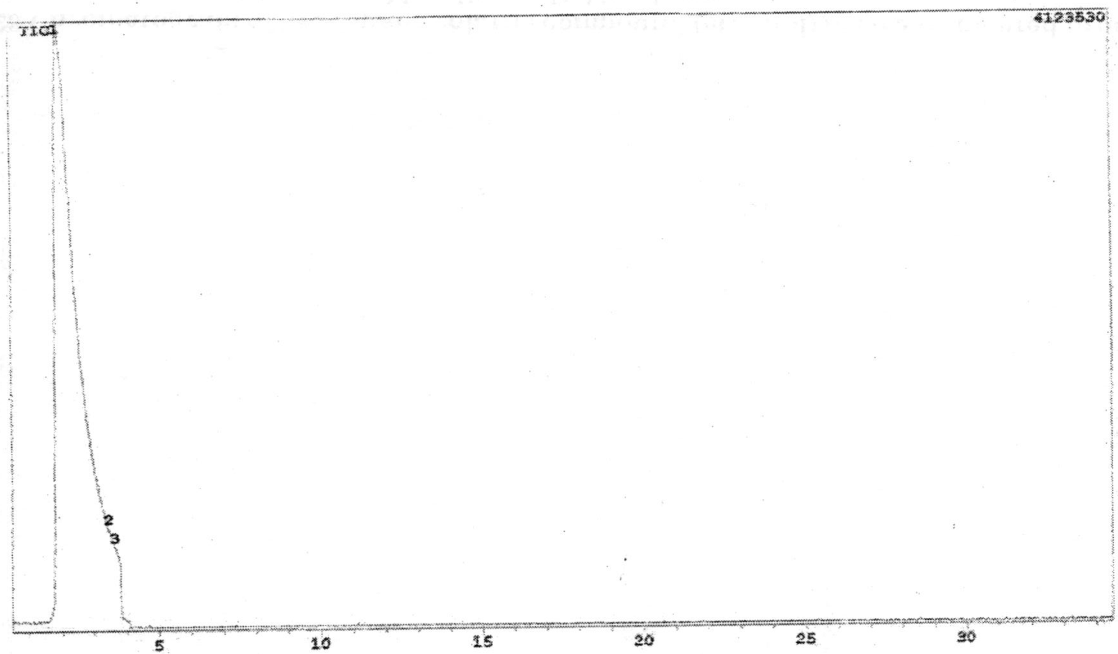

Fig. 22 Chromatogram of headspace volatile compound (ethanol) of rice fermented product by *M. Indicust + C. Parpsilosis*

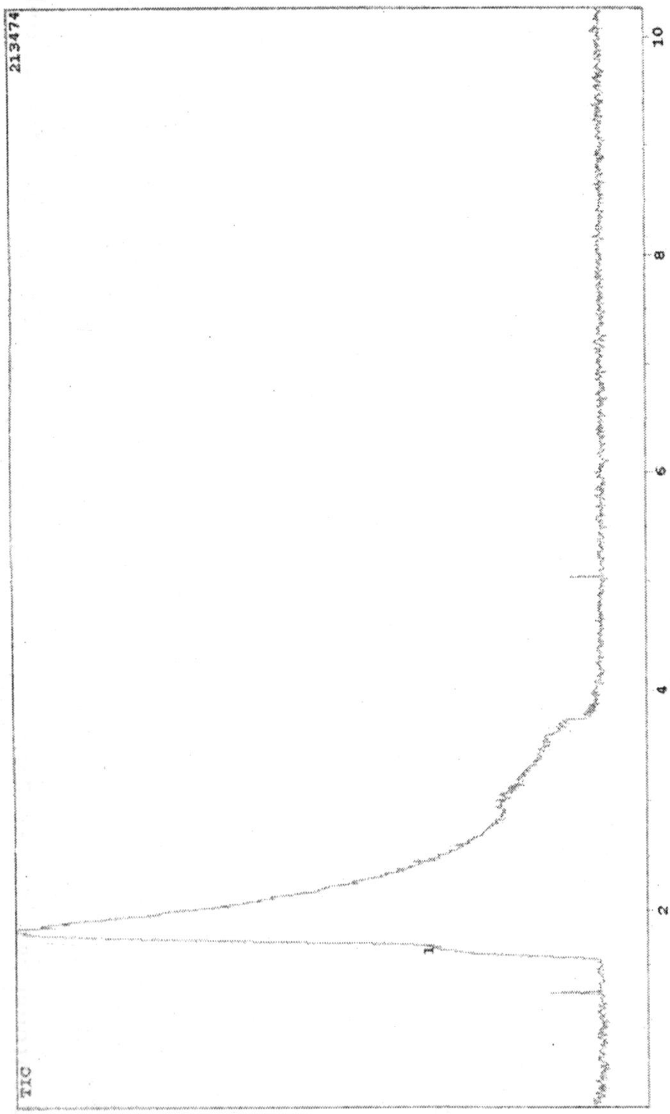

Fig. 23 Chromatogram of headspace volatile compound (ethanol) of rice fermented product by *M.Indicus* + *R. Mucilaginosa*

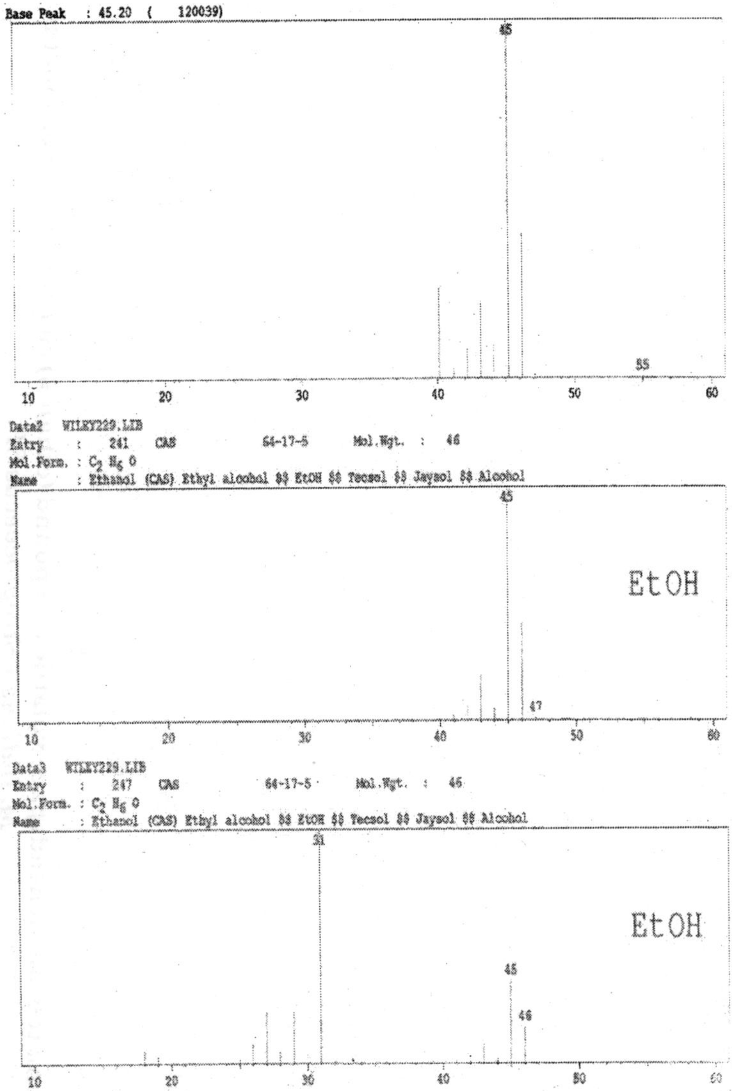

Fig. 24. GCMS profile of Volatile compound (ethanol) of rice fermented product by *M. Indicus* + *C. Parapsilosis* and *M. Indicus* + *R. mucilaginosa*

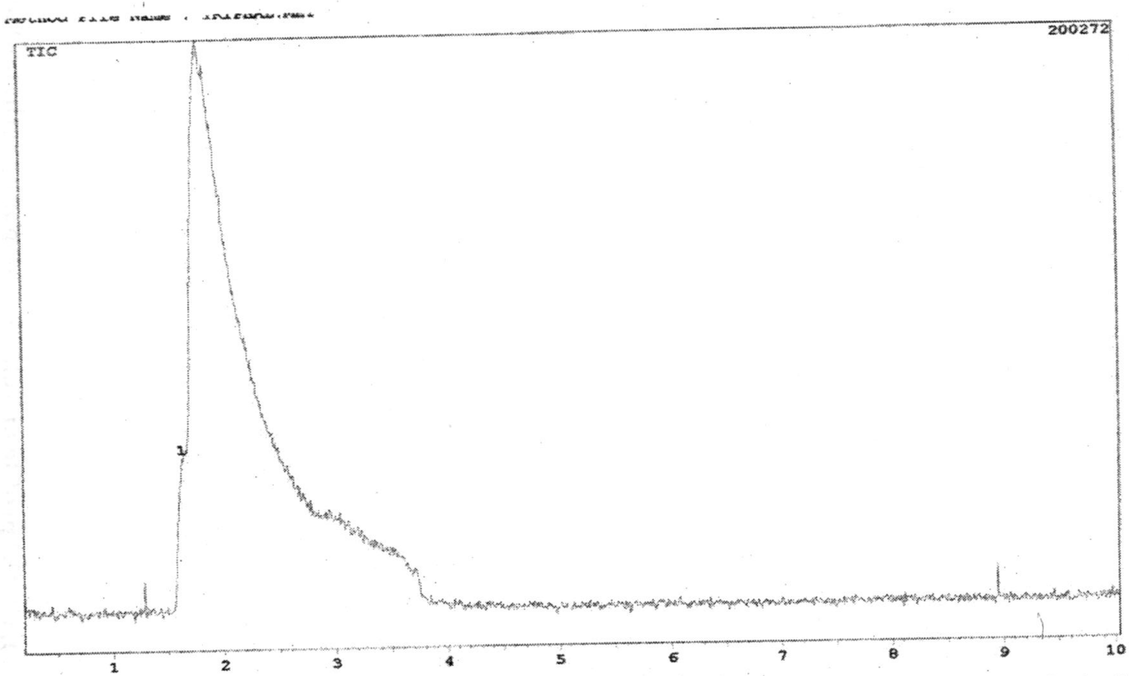

Fig. 25 Chromatogram of headspace volatile compound (ethanol) rice fermented product by *M. Indicus + Streptococcus SP.*

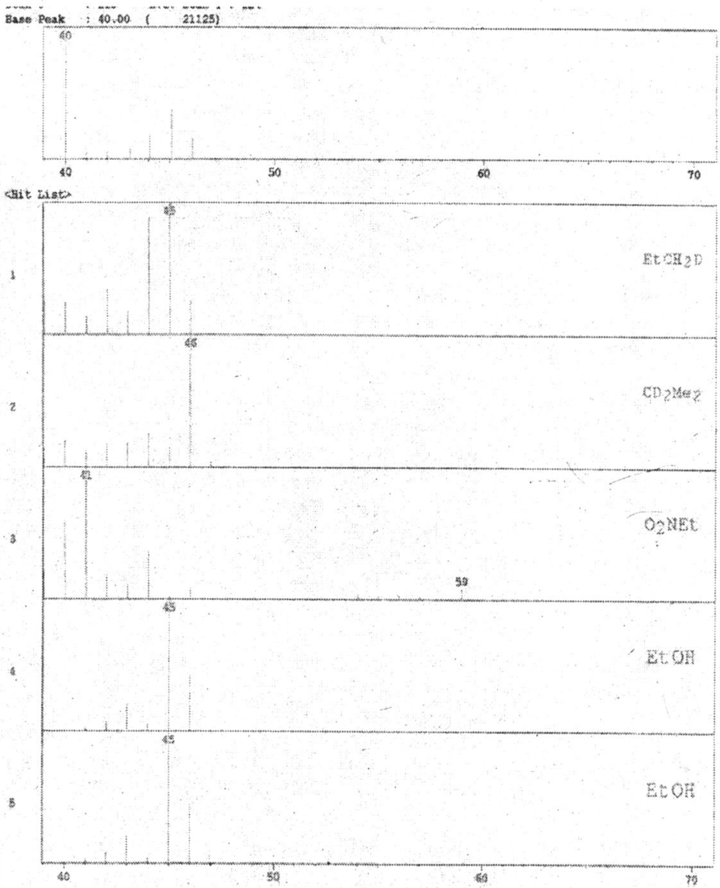

No	SI	Mol.Wgt.	Mol.Form./Compound Name	CAS No.	Entry	LIB#
1	73	44	C_3H_7D 1-DEUTEROPROPANE $$	20717-73-1	198	2
2	71	44	$C_3H_6D_2$ 2,2-DIDEUTEROPROPANE $$ DEUTEROPROPANE $$	2875-95-8	191	2
3	71	75	$C_2H_5NO_2$ Ethane, nitro- (CAS) Nitroethane $$	79-24-3	1593	2
4	70	46	C_2H_6O Ethanol (CAS) Ethyl alcohol $$ EtOH $$ Tecsol $$ Jaysol $$ Alcohol $$ Algrain $$ Anhy	64-17-5	241	2
5	70	46	C_2H_6O Ethanol (CAS) Ethyl alcohol $$ EtOH $$ Tecsol $$ Jaysol $$ Alcohol $$ Algrain $$ Anhy	64-17-5	247	2

Library Name
(1) SZTERP.LIB (2) WILEY229.LIB

Fig. 26 GCMS profile of Volatile compound (ethanol) of rice fermented product by *M. Indicus* + *Streptococcus SP.*

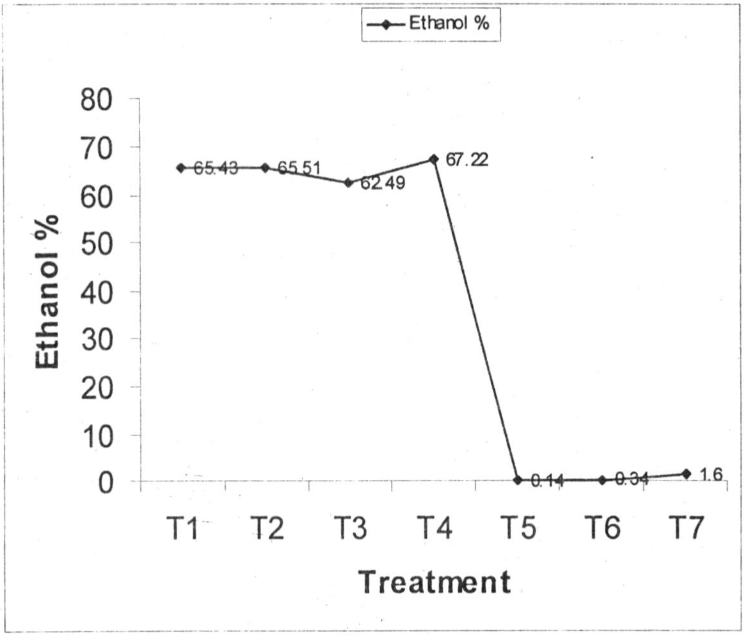

Fig. 27: Ethanol yield from Hamei of five isolated microorganisms (mould, yeast& bacteria)

Fig. 28.On critical analysis of the results it was reveals that ***hamei*** made by adding *A. kalkora* bark produces more or less similar ethanol yield with ***hamei*** made by adding *A. myriophylla* (*which* is found to be 76.51%). Water quantity of 1:5 w/v at the time of cooking rice produces more yield than 1:10 w/v water volume. Inoculum in the form of ***hamei*** is found to be more effective than 1 ml suspension of mould culture grown overnight; however when it is supplemented by bark powder in the fermented broth, the yield increases. Data indicated that in 1:5 w/v water volume at the time of cooking, ***hamei*** of *A. myriophylla* bark powder produces 75.62%, ***hamei*** of *A. kalkora* bark powder produces 76.51%, 1 ml culture suspension produces 60.29%, 1 ml culture suspension + *A. myriophylla* bark produce 70.18%, 1 ml suspension + *A. kalkora* bark produce 70.98%, whereas in 1:10 w/v water volume, it produces 63.65%, 65.28%, 54.98%,

58.59% and 59.05% respectively

4.5.2.0 Bark powder percentage in *Hamei*

Effects of bark powder percentage in ***hamei*** on ethanol yield are presented in Table XI and graphically depicted in Fig. 29. When the data is critically analysed, significant difference in yield with respect to bark percentage is observed. Maximum yield is recorded from 2% concentration as 76.89% and minimum is recorded from 5% concentration as 69.52, while for concentration of 1% and 3% the yield recorded is at 73.38% and 73.95% respectively.

4.5.2.1. Incubation period

Data for ethanol yield with respect to incubation period are depicted in Table XII and graphically depicted in Fig. 30. A critical examination of the data reveals that there are significant variations of ethanol yield in different incubation period of ***hamei*** and fermentation broth. Maximum yields are recorded in 120 hours, which was 77.35% and 76.27% for ***hamei*** and fermentation broth respectively. On further increase of incubation time to 144 hours the yield is found to decrease to 73.61% for ***hamei*** and 72.33% for fermentation broth.

Table X

Effect of *Albizia* variety in Hamei, Inoculum type and Water quantity (1 : 5 w/v and 1:10 w/v) in fermentation on ethanol yield and (1 : 10) w/v in fermentation

Treament	*Water ratio in mash (w/v)*	
	1:5	*1:10*
Albizia myriophylla bark Hamei	75.62 ± 0.11	63.65 ± 0.51
Albizia kalkora bark Hamei	76.51 ± 0.45	65.28 ± 0.54
1ml mould culture suspension	60.29 ± 1.00	54.98 ± 0.21

(Contd...)

1ml mould culture suspension+ *A.myriophylla* powder	70.18 ± 0.60	59.59 ± 0.26
1ml mould culture suspension+ *A.kalkora* bark powder	70.98 ± 0.13	59.05 ± 0.35
Control (Hamei without bark powder)	65.30 ± 1.00	60.81 ± 0.51
CD at 5%	SE = 0.46 CD = 0.99	SE = 0.20 CD = 0.58

Table XI
Effect of bark powder quantity in Hamei

ETHANOL % (v/v)				
Bark powder % in HameiTreatment	*1*	*2*	*3*	*5*
Albizia Kalkora	73.38 + 0.10	76.89 + 0.01	73.95 + 0.04	69.52 + 0.02
CD at 5%		SE = 0.02	CD = 0.07	

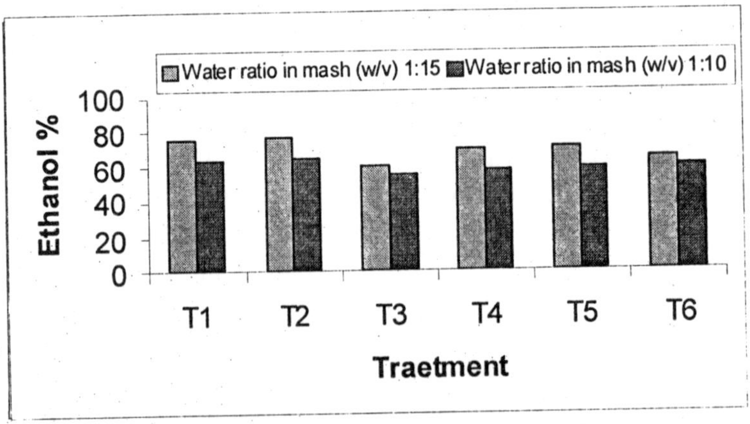

Fig.28 : Effect of *Albizia* variety in Hamei, inoculum type and water quantity (1:5 w/v and 1:10 w/v) in fermentation on ethanol yield

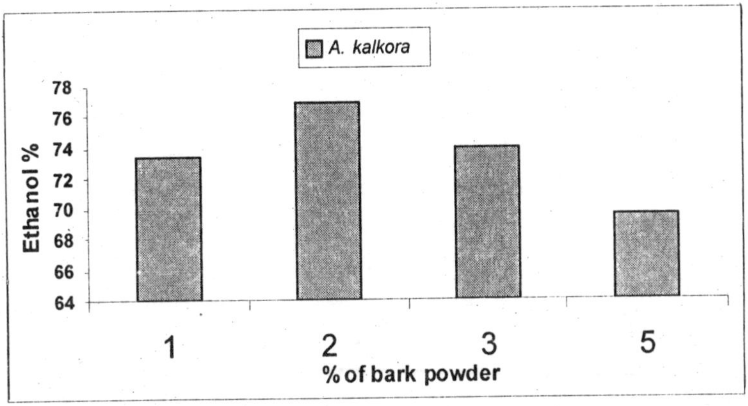

Fig. 29: Effect of bark powder quantity in Hamei

4.5.2.4. Rice quality and quantity of Hamei inoculum

Data presented in Table XII and result depicted graphically in Fig. 31 clearly indicate that the effect of the rice quality in ***hamei*** and fermentation broth and quantity of ***hamei*** inoculum on ethanol yield is highly significant. Maximum ethanol is recorded from 12 gm of unpolished rice (local) ***hamei*** and 50 gm unpolished rice fermentation broth as 75.68% and minimum yields are recorded from 6 gm polished rice ***hamei*** and polished rice fermentation broth as 66.71%. It is observed that when only 6 gm of ***hamei*** is inoculated, the ethanol yield decreases, but local variety (unpolished rice) ***hamei*** and fermentation broth produce better yield than the market rice (milled). Data reveal that 12 gm ***hamei*** *and* 6 gm ***hamei*** inoculated local rice fermentation broth (50 gm cooked rice) produce 75.68% and 73.76% of ethanol respectively; but when market rice is used in the fermentation broth, it produces 70.09% and 69.79% respectively. When 12 gm and 6 gm of market rice ***hamei*** are inoculated to fermentation broth of local rice it produces the yield of 71.74% and 70.33% and in market rice fermentation broth it produces 67.64% and 66.71%.

4.5.2.5. Comparative ethanol yield from hamei and fermentation broth of different rice cultivars

Data of ethanol yield presented in Table XIV and graphically depicted in Fig.32 reveals that the yield in respect of different rice cultivars in ***hamei*** and fermentation broth are highly significant. A critical examination of the result reveals that there are significant variations of ethanol yield among different rice grains. Maximum ethanol yields are recorded as 75.29% from RCM-9 (T_{14}) followed by 74.62% from KD-2-6-3 (T_{10}) and minimum yields are recorded as 65.55% from IR-64(T_{13}) in the case of ***hamei***. In fermentation broth maximum yields are recorded as 76.22% from RCM-5 (T_{14}) followed by 75.53%, 75.21%, 74.55% and 73.54% from RCM-9 (T_{15}) RCM-10 (T_{16}), KD-2-6-3(T_{10}) and **De se** (T_9) respectively and the minimum is recorded as 68.38% from **Ritiang**(T_5)

Table XII
Effect of incubation period of Hamei and fermentation broth on ethanol yield.

Period of incubation / Treatment	*Ethanol % (v/v)*					
	24 hrs	*48 hrs*	*72 hrs*	*96 hrs*	*120 hrs*	*144 hrs*
Hamei	38.06± 0.88	52.88± 0.85	69.99± 1.82	72.40± 0.62	77.35± 0.70	73.61± 0.59
Fermentation Broth	46.74+ 1.26	53.66+ 0.85	69.91+ 1.22	73.44+ 0.83	76.27+ 1.22	72.33+ 0.55

CD at 5% SE = 0.91 CD = 1.94

Table XIII
Effect of rice quality and quantity of Hamei inoculum on ethanol yield

Fermentation broth	*Ethanol % v/v*	
	Polished rice	*Unpolished*
Hamei	*(50 gm)rice*	*(50 gm)*
12 gm Hamei unpolished rice	70.09 ± 0.68	75.68±0.72
!2 gm Hamei polished rice	66.71± 0.31	71.74±0.81
6 gm of unpolished rice	69.79 ± 1.75	73.76±0.51
6gm of polished rice	76.64 ± 0.81	70.33±1.48
CD at 5%	SE = 0.87	CD = 1.85

Table XIV
Ethanol Yield from different Rice cultivars in Hamei and Fermentation broth

Treatment	*Ethanol% (v/v)*	
	Hamei	*Fermented broth*
T_1 MPANGBAN	70.47 ± 0.36	69.59 ± 1.66
T_2 TUNANAPNANG	70.45 ± 0.29	69.44 ± 0.87
T_3 NAPDAI	70.46 ± 1.69	71.79 ± 1.47
T_4 BANGLAI	68.11 ± 0.13	70.06 ± 1.13
T_5 RITIANG	69.27 ± 0.67	68.38 ± 0.41
T_6 NAPPOK	68.24 ± 1.03	71.46 ± 0.30
T_7 SHAJEI	70.33 ± 0.26	72.58 ± 0.24
T_8 SANGBU	69.64 ± 1.51	68.88 ± 0.26
T_9 DE SE	73.54 ± 0.33	74.55 ± 0.26
T_{10} KD_{2-6-3}	74.62 ± 0.29	75.21 ± 0.52
T_{11} TAMPHAPHOU	68.5 ± 1.26	71.75 ± 0.85
T_{12} CAU R_3	69.69 ± 0.33	71.87 ± 0.55
T_{13} IR_{64}	65.55 ± 0.23	69.19 ± 1.22
T_{14} RCM_5	73.34 ± 0.55	76.22 ± 0.52
T_{15} RCM_9	75.29 ± 1.74	75.58 ± 0.26
T_{16} RCM_{10}	72.85 ± 0.76	75.53 ± 0.38
CD at 5%	SE =0.76 CD = 1.62	SE = 0.63 CD = 1.35

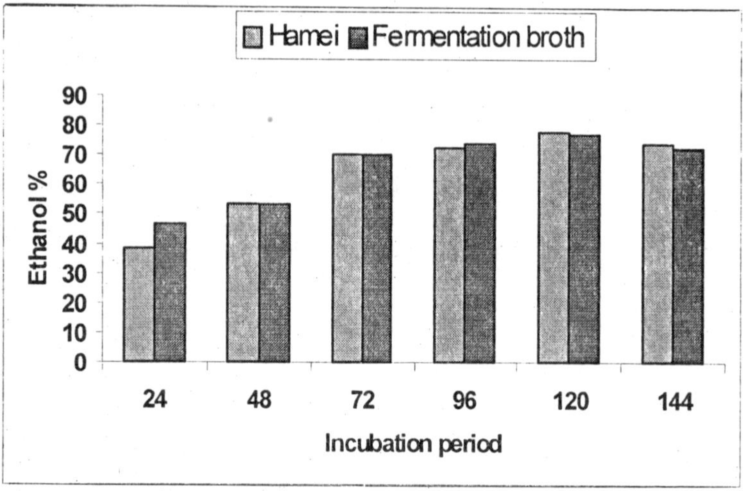

Fig. 30: Effect of incubation of Hamei and fermentation broth on ethanol yield

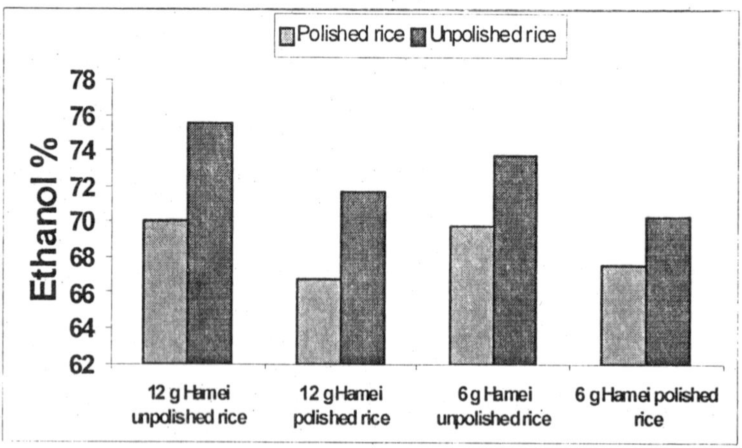

Fig. 31: Effect of rice quality and quantity of Hamei on ethanol yield

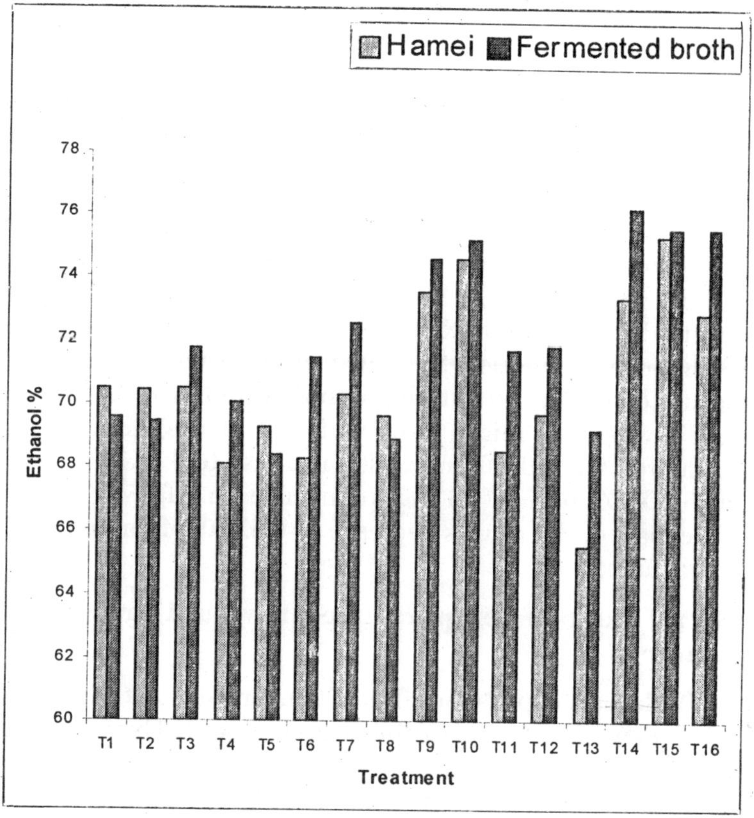

Fig. 32: Ethanol yield from differnt rice cultivars in Hamei and fermentation broth

4.5.2.6 Phenolic acid in hamei

Table XV and Fig.33 which graphically depict the results of ethanol yield from ***hamei*** prepared by 2% caffeic acid, 2% ferulic acid, 2% vanillic acid, 2% salicylic acid and mixture of four acids reveals that ***hamei*** of acid mixture yields 76.32% ethanol which is as high as the controlled (***Hamei*** of *Albizia* bark) which is 75.09%; yields from other individual acids are 53.86%, 51.17%, 53.02 % and 52.85% respectively. Maximum yields are recorded from ***hamei*** of acid mixture and minimum from 2% ferulic acid.

4.6.1. Growth Kinetics of *M.indicus* in *hamei* and fermentation broth (CFU/ml

Data recorded in Table XVI and Fig. 34 (graphical representation) reveal the growth kinetics of *M.indicus* in *hamei* and fermentation broth in respect of incubation period. Counts in *hamei* with bark powder and without bark powder are found to be $2.95X10^6$ cfu/g and $1.2X10^6$ cfu/g at the end of 24 hours incubation and increase to $6.2X10^6$ cfu/g and $5.7X10^6$ cfu/g after 96 hours of incubation. However the counts decrease to $5.3X10^6$ cfu/g and $4.45X10^6$ cfu/g after 120 hours of incubation. Similarly, in the fermentation broth inoculated by *hamei* with bark powder and *hamei* without bark powder, the initial counts (after 24 hours) are $2.3X10^6$ cfu/ml and $2.2X10^6$ cfu/ml, which increase to $6.6X10^6$ cfu/ml and $5.8\ X10^6$ cfu/ml at 96 hours of incubation. After 120 hours of incubation the counts decrease to $6.3X10^6$ cfu/ml and $5.3\ X\ 10^6$ cfu/ml.

4.6.2. Effect of temperature on visual growth of *M.indicus*

Table XVII depicts the effect of temperature on the growth of *M.indicus*. Temperature tolerance is observed between the ranges of 15 to 37°C. Organisms are not able to grow at 10°C and below, while growth is found to be poor at 15°C, at 40°C and above, organisms are not able to grow since they lose viability.

Table XV

Effect of phenolic acids in Hamei on Ethanol yield

Treatment T	*Ethanol % (v/v)*
T_1 2% caffeic	56.86 ± 0.72
T_2 2% ferulic acid	51.17=0.48
T_3 2% salicylic	53.02=0.55
T_4 2% Vanillic acid	52.85=1.02
T_5 mixed acid	76.32=0.55
T_6 control (Hamei of A.kalkora)	75.09=1.44
CD at 5%	SE = 0.74 CD = 1.58

Table XVI
Growth kinetic of *M. indicus* in Hamei and fermentation broth with respect to incubation time

Incubation period / *Treatment*	*24 hrs*	*48 hrs*	*72 hrs*	*96 hrs*	*120 hrs*
Hamei without Albizia bark powder	1.2 x 10^6	2.35x10^6	3.35x10^6	5.7x10^6	4.45x10^6
Hamei with Albizia bark powder	2.29x10^6	3.65x10^6	4.7x10^6	6.2x10^6	5.3x10^6
Broth of hamei without Albizia bark powder	2.2x10^6	4.1x10^6	5x10^6	5.8x10^6	5.3x10^6
Broth of hamei with Albizia bark powder	2.3x10^6	4.7x10^6	5.9x10^6	6.6x10^6	5.3x10^6

Table XVII
Effect of temperature on visual growth of *M. indicus* Treatment Period

Temperature	*O/N*	*48 hrs*
4°C	–	–
10°C	–	–
15°C	T	+
26°C	++–	++++
30°C	++–	++++
37°C	++–	++–
40°C	–	–
45°C	–	–
60°C	–	–

Symbols

	no visible growth
T	trace of visible growth
+	scattered colonies
+ +	mycelium growth almost covered the surface
+ + -	luxuriant growth covered the entire surface
+ + + +	heavy growth filled the flask

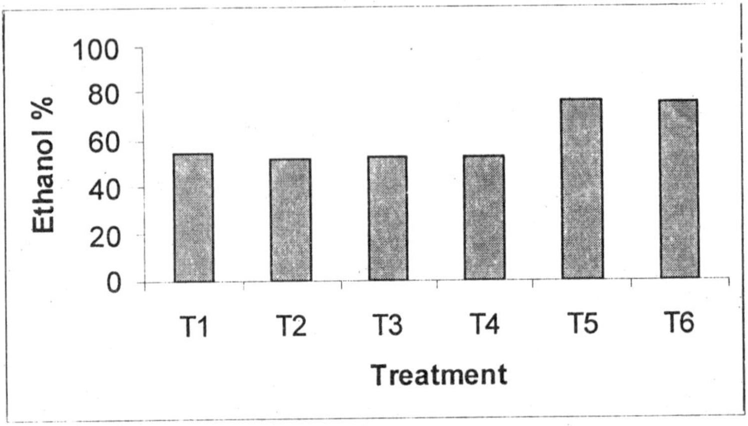

Fig. 33: Effect of phenolic acids in Hamei on ethanol yield

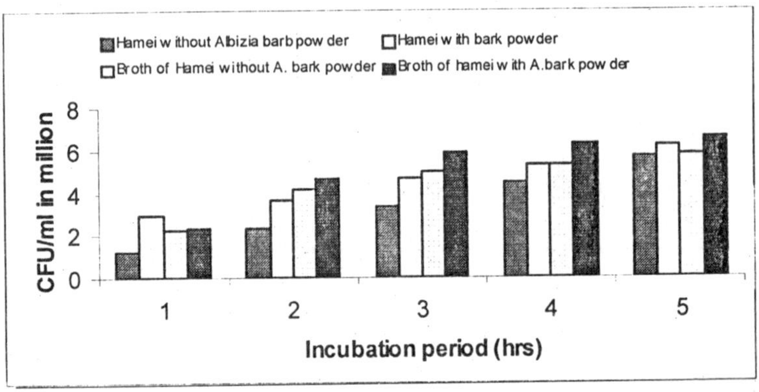

Fig.34: Growth kinetic of *M. indicus* in Hamei and fermentation broth with respect to incubation time

4.6.1 Effect of pH on visual growth of *M.indicus*

pH effect on visual growth of *M.indicus* is presented on Table XVIII. It is found that organisms grow well between the pH range of 5 to 9. At pH 10, growth is found to be slow and at 4 and 3 growth is observed after 48 hours and 72 hours of incubation. Viability is observed even at pH 2 however it is not able to grow at this pH.

4.6.2 Effect of ethanol on visual growth of *M.indicus*

Data presented in Table XIX and Figs. 35 and 36 depict the ethanol tolerance of *M. indicus* with respect to different period of incubation time and different concentration of ethanol. From the observation it is found to be completely tolerant up to 25% concentration but decrease in colony count (cfu) is observed as the treatment duration increases. At 20% concentration, treatment after 9 hours is found to be lethal. At 25% concentration a few viable colonies are found at 2 hours; treatment beyond this duration resulted in complete loss of cell viability. It is found to be completely tolerant at 15% when treated for 24 hours.

Table XVIII

Effect of pH on visual growth of *M. indicus*

pH	*O/N*	*48 hrs*
2	–	–
3	–	–
4	T	T
5	++–	++++
6	++–	++++
7	++	++++
8	++++	++++
10	+	+

Symbols :

-	no visible growth
T	trace of visible growth
+	scattered colonies
+ +	mycelian growth almost covered the surface
+ + -	luxurian growth covered the entire surface
+ + + +	heavy growth filled the flask

Table XIX
Ethanol Tolerance of *M. indicus*

	Colony forming Unit (CFU / ml.)					
Ethanol % / *Treated Period*	*0*	*1*	*5*	*10*	*15*	*20*
2hrs	$1.3x\ 10^7$	$1x\ 10^7$	$6x10^6$	$3.8x10^6$	$1.8x10^6$	$6.6x10^5$
6hrs	$1.32x\ 10^7$	$9.6x\ 10^6$	$5.2x10^6$	$3.4x10^6$	$1.2x10^6$	$1x\ 10^3$
9hrs	$1.32x\ 10^7$	$9x\ 10^6$	$4.2x10^6$	$2.8x10^6$	$6x10^4$	$2x10^2$
15hrs	$1.35x10^7$	$8x\ 10^6$	$3.6x10^6$	$2.2x10^6$	$8x10^4$	_
24hrs	$1.37x10^7$	$1.2x\ 10^6$	$1.16x10^6$	$1.04x10^5$	$1x10^7$	_

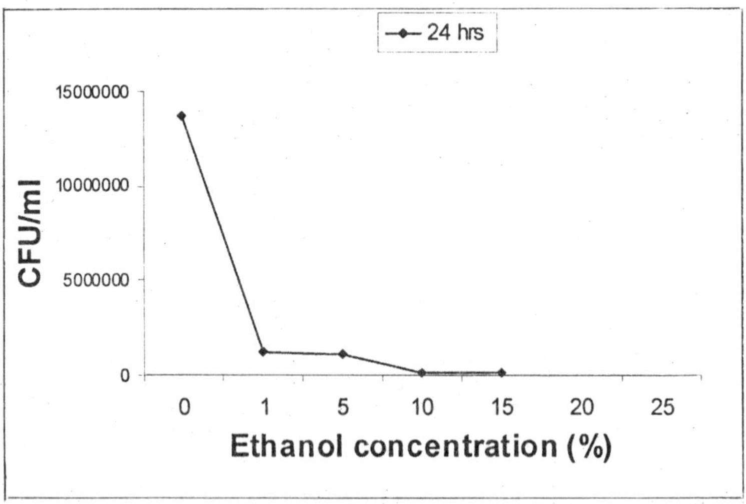

Fig. 35: Ethanol tolerance of *M. indicus* when treated for 24 hours with different ethanol concentration

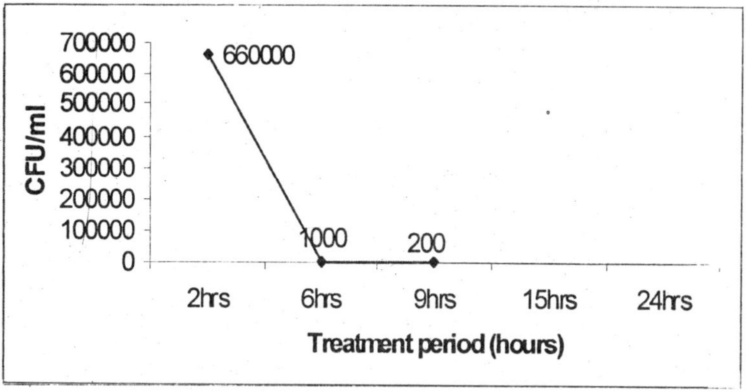

Fig. 36: Ethanol tolerance of *M. indicus* when treated with 20% ethanol for different time period

COLOUR PLATES 17-32

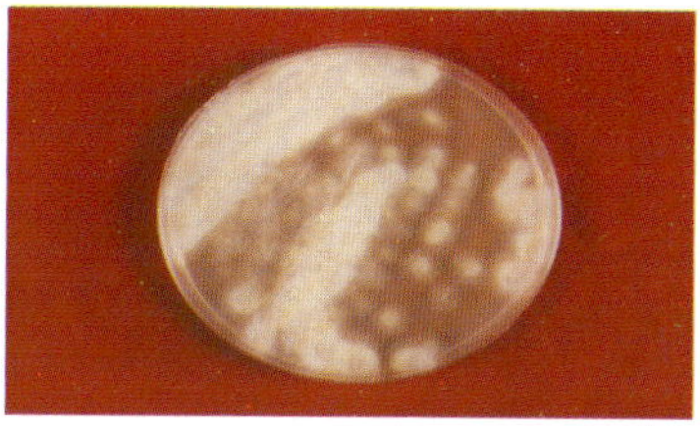

17. Colony of *Mucor indicus* in YPDA plate

Plate 18. *Mucor indicus* (mycelium) under microscope

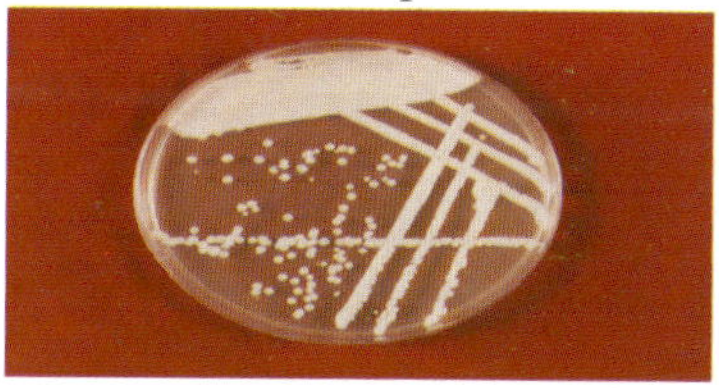

Plate 19. Colony of *Candida parapsilosis* in YPDA plate

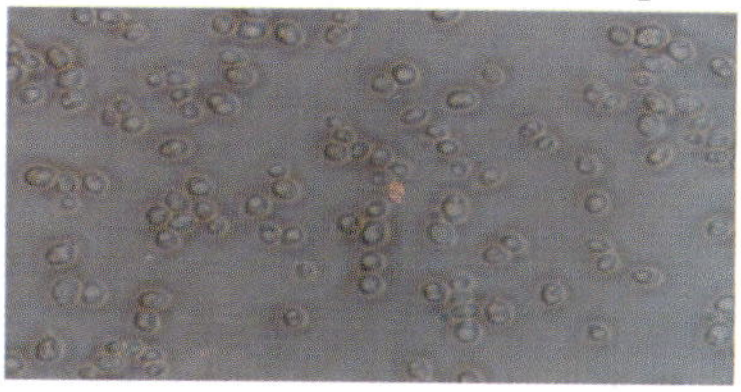

Plate 20. *Candida parapsilosis* under microscope

Plate 21. Colony of *Rhodotorula mucilaginosa* in YPDA plate

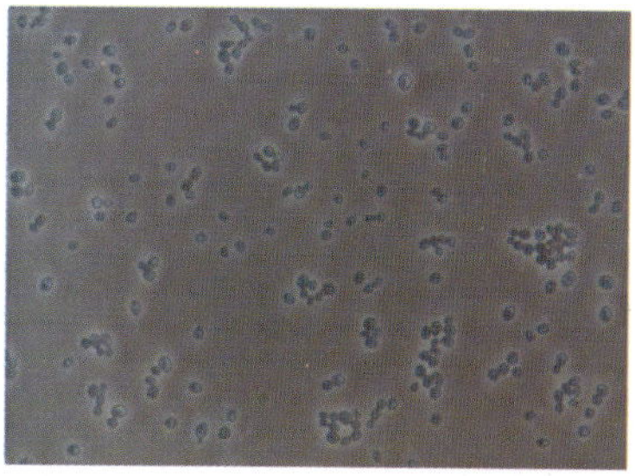

Plate 22. *Rhodotorula mucilaginosa* under microscope

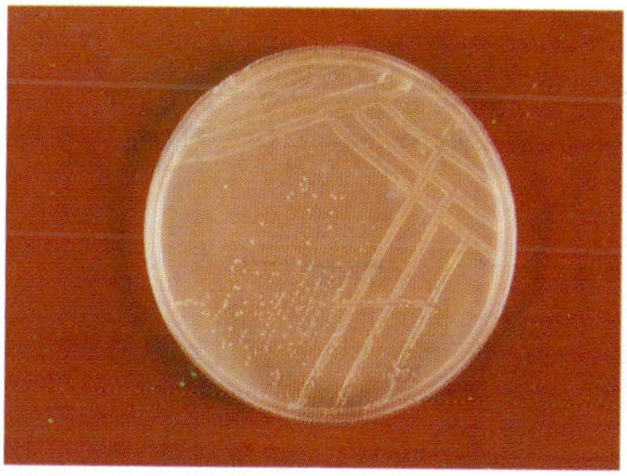

Plate 23. Colony of *Streptococcus sp.* in YPDA plate

Plate 24. *Streptococcus sp.* under microscope

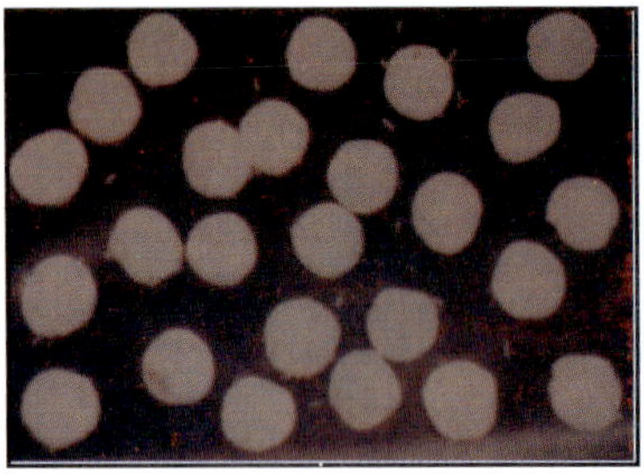

Plate 25. Hamei cake (Traditional

Plate 26. Hamei cake of *Mucor indicus* (Laboratory)

Plate 27. Fermentation broth (traditional)

Plate 28. Fermentation broth (Laboratory)

Plate 29. Traditional distillation of 'Yu'

Plate 30. Colony morphology of *Mucor indicus*

5

Discussion

The present investigation on the "Role of *Albizia spp.* in the fermentation process of different rice cultivars" is discussed in the light of the findings as recorded in the preceding chapters. The results show that *Albizia* bark powders contribute a beneficial role in the alcoholic fermentation of rice. The contributing characters to alcohol production and chemical composition of substrate (*Albizia* bark powder and Rice) are studied.

Albizia plants whose bark powders are used in the preparation of ***hamei*** are identified as *A.Myriophylla* Benth. and *A.Kalkora* Pain according to Kanjilal and Das (1939). *Albizia myriophylla* is a large woody climbers whereas *Albizia kalkora* and erect tree. The fibrous bark is sweetish and astringent in taste. It may be due to the presence of phenolic compounds. Ito et al. (1995) reported similar findings that *Albizia myriophylla* has so far been the only plant that is known to have sweet taste among *Albizia* species and isolated four lignan glycosides from the bark.

Based on personal experiences of the brewers, rice variety or cultivars, which are easily and commonly available and generally used were selected. Some glutinous rice, semi-glutinous, non-glutinous, medium grain and short grains are used for the fermentation. Yoshizawa and Kodama (1977) also reported that rice of short grain varieties are considered best for *sake* manufacture and large grains were also considered desirable.

5.1.1. Moisture content

Moisture content is an important physical parameter in determining the quality of grain, stability and economic value. There are significant variations of moisture content among 16 rice cultivars (Table 1 and Fig5). Significantly higher moisture percentage is observed in **Mpangban** {(T_1) 11.59%) followed by **Tunanapnang** {(T_2) 11.32%} and **Napdai** {(T_3) 11%}. The minimum moisture content is observed in **Ritiang** {(T_5) 7.41%} followed by **De se** {(T_9) 7.45%} and **Sangbu** {(T_8) 7.83}. The findings show that moisture contents of rice grain are in the range of 7.4 to 11.59%, which is within the tolerance limit. Variation in moisture content could be attributed to the changes in ambient temperature and relative humidity. The findings of the present investigations can be supported on the basis of findings of Dhaliwal et al. (1991) who pointed out that moisture content of stored paddy equilibrates according to the temperature and relative humidity regardless of initial moisture content. Similar findings are reported by Deka et al. (2000) that moisture percentage of **Basmati** rice *Oryzae sativa L.* ranges from 11.15% to 11.29%. Physical and biochemical characteristics of grains depend upon moisture content (Biskupski et al., 1979). In rice, starch granules are fairly hygroscopic; therefore the moisture content of rice grain readily changes with the change of storage temperature and relative humidity (Araullo et al., 1976). At increasing moisture levels, a high correlation is found between increase in mould growth, carbon dioxide production, reducing sugar, decrease in germinating capacity and non-reducing sugar (Lund et al., 1971).

5.1.2. Ash content

Ash is an index of mineral content of grain. Ash contains

several nutritionally important minerals in varying proportion. Minerals make up the major part of inorganic substance to form vital molecules in the grain. The ash content of cereal grain can be regarded as indicator of quality of whole grain.

In the present investigation, highest levels of ash content (Table 1 and Fig.5) are observed in KD-2-6-3 (T_{10}) 1.96%}, followed by RCM-10 {(T_{16}) 1.93%}, IR-64 {(T_{13}) 1.90%} and **Tunanapnang** {(T_2) 1.88%}. Minimum levels are observed in **De se** {(T_9) 0.96%} and **Sangbu** {(T_8) 1.16%}. Ash content for the rest of samples lies between the ranges of 0.96% to 1.96%. Similar findings of ash content in twenty rice varieties of Northern India are in the range of 0.9% to 1.41% as reported by Raghaviah and Kaul (1970). Juliano et al. (1964) also observed that the variation of ash content in brown and polished rice is from 1.53 to 2.13 percent and 0.52 to 0.74 percent respectively.

5.2.1. Starch content

Starch is the principle carbohydrate of rice and a major source of reserve energy used to carry out life processes. Different rice cultivars have a significant variation of starch content (Table II and Fig.6). Highest starch content is recorded in **Mpangban** {(T_1) 88.48 mg/g} followed by **Tunanapnang** {(T_2) 87.95 mg/g} **Napdai** {(T_3) 79.78 mg/g} and KD-2-6-3 {(T_{10}) 78.43 mg/g}. This is expected because all these varieties are gluntinuous rice, which are not favourable for eating as a regular food. A similar findings are reported by Dutta and Baruah (1978) that the starch content of some glutinuous and non-glutinuous rice varieties of Assam ranges from 73.44 to 81.22%. Little variation of starch content are observed in RCM-10 {(T_{16}) 76.40 mg/g}, RCM-9 {(T_{15}) 76.18 mg/g} and RCM-5 {(T_{14}) 76.16 mg/g}. A minimum content is observed in IR-64 {(T_{13}) 69.89}. The findings are in conformity with the findings of Firmin (1991) that a variation of starch content ranges from 67.7 to 75.5%. Chakravorty et al. (1972) reported that starch which constitutes the bulk of rice endosperm is found in the range of 79.2 to 81.4% of the grain. Al-Bayati and AI Rayes (1981) also reported the variation of starch content in **Iragi** brown rice from 83.62 to 87.99% respectively.

5.2.2. Amylose content

Starch is made up of glucose units of two distinct fractions: a linear fraction amylose and a branched fraction amylopectin. The amylose/amylopectin ratio is the major fraction influencing the water absorption and volume expansion during cooking, as well as texture and gloss of boiled and milled rice (Araullo et al., 1976).

The present investigation for amylose content of 16 different rice cultivars (Table II and Fig. 6) shows that IR-64 {(T_{13}) 26.87 mg/g} contains the maximum quantity followed by **Banglai** {(T_4) 26.71 mg/g,} **De se** {(T_9) 26.59 mg/g}, **Ritiang** {(T_5) 26.18 mg/g} and **Sangbu** {(T_8) 26.08 mg/g}. The findings are in conformity with the finding of Bhattacharya et al. (1980), who reported that amylose contents of 100 varieties of rice samples collected from different region of India are as high as 26%. Amylose contents from 22.4 to 27.9 percent in certain long grain rice are also reported by Wardworth and Spadaro (1981). However, minimum amylose content is observed in **Mpangban** {(T_1) 10.11 mg/g} followed by **Tunanapnang** {(T_2) 14.28 mg/g}, KD-2-6-3 {(T_{10}) 15.73 mg/g} and **Napdai** {(T_3) 19.75 mg/g}. Similar findings are reported by Raghaviah and Kaul (1970) that amylose content of some high yielding varieties of India varies from 16.5 to 35 percent. Juliano et al. (1964) also reported that amylose content varies from 13.7 to 32.8 percent in few rice varieties grown in some Asian countries. Sharma and Baruah (1979) also observed that average amylose content of brown rice ranges from 15.4 to 16.9 percent. Another report given by Dutta and Baruah (1978) shows that amylose content of some glutinuous and non-glutinuous rice varieties grown in Assam ranges from 17.14 to 31.97 percent respectively.

5.2.3 Amylopectin content

Amylopectin molecules are the major constituents of starch grain and much larger than those of amylose with multiple branches.

In the present investigation amylopectin content differs significantly among 16 rice cultivars (Table II Fig.6). Significantly high content of amylopectin is recorded in **Mpangban** {(T_1) 89.89 mg/g} followed by **Tunanapnang** {(T_2) 85.71}, KD-2-6-3 {(T_{10}) 84.27 mg/g} and **Napdai** {(T_3) 80.25 mg/g}. The minimum is recorded in IR-64{(T_{13}) 73.13 mg/g}. The result is expected because most of the rice cultivars used in the experiment are glutinous rice. Similar findings are reported by Singh (1989) that waxy or glutinuous varieties of rice produce starch comprising almost entirely of amylopectin, (75-85 percent).

5.3.1. Total soluble sugar

Rice contains several sugars in free state which are known as total soluble sugars. Sugars are the simplest form of carbohydrate; some of them are reducing while few of them are non-reducing in characters.

There are significant variations of total soluble sugar content among rice grains (Table III and Fig.7). Maximum total soluble sugar content is observed in RCM-9 {(T_{14}) 1.17 mg/g} followed by **Mpangban** {(T_1) 1.16 mg/g}. However the variation found is insignificant among other varieties viz. **Tananapnang** {(T_2) 1.04mg/g} and **Napdai** {(T_3) 1.04mg/g}. Minimum contents are observed in **Sangbu** {(T_8) 0.66 mg/g} and IR-64 {(T_{13}) 0.68 mg/g}. Overall sugar contents among the rice grains are found at the range of 1.16 mg/g to 0.66 mg/g. The present findings are in conformity with the findings of Kondo and Okamura (1934) and Williams and Bevenue (1953).

5.3.2. Reducing sugar

Rice mainly contains reducing sugar such as glucose, fructose and maltose. Significant differences of reducing sugar level are observed among different rice cultivars (Table III and Fig. 7). Insignificant content of reducing sugars are found in cultivars like **Nappuak** (T_6), ***Sangbu*** (T_8), **De se** (T_9) and CAU-3(T_{12}) since they have similar value of 0.19 mg/g. Similarly, the same value is obtained in RCM-5(T_{14}) and RCM-9(T_{15}) at 0.28 mg/g. Maximum reducing sugar contents are recorded in **Mpangban** {(T_1) 0.52 mg/g}, followed by **Tunanapnang** {(T_2) 0.47 mg/g}, KD-2-6-3 (T_{10}) 0.42 mg/g}, **Banglai** {(T_4) 0.38 mg/g} and **Napdai** {(T_3) 0.37 mg/g}. This is expected because these rice cultivars are sweet and glutinuous. This could be explained due to higher degradation of starch by the starch hydrolysing enzyme viz. á and â amylases. The findings are in conformity with the findings of Ali (1997). Minimum content is observed in IR-64 {(T_{13}) 0.14 mg/g}. Similar findings are reported by Sinha and Narasimhan (1967) and Pushpama and Reddy (1979).

5.3.3. Non-Reducing sugars

The non-reducing sugars are mainly sucrose and traces of raffinose, and maltotriose. These sugars improve digestibility and provide better taste in rice.

Significant differences are observed among 16 rice cultivars in respect of non-reducing sugar (Table III and Fig. 7). Maximum value is recorded in RCM-9 {(T_{16}) 0.89 mg/g} followed by RCM-5 (T_{34}) 0.85 mg/g} and RCM-10 {(T_{16}) 0.80 mg/g}; similar observations are reported by Sreenivasan (1939), Lynch et al. (1962), Huang (1986) and Narayan et al. (1988). Minimum reducing sugar content is observed in **Sangbu** {(T_8) 0.47 mg/g}. Similar observations are made by

Mathur and Sinha (1978). Non-reducing sugar contents are recorded in the range between 0.89 mg/g and 0.47 mg/g among 16 rice cultivars. A wide variation in the non-reducing sugar content of rice varieties and also between the rough rice (paddy), brown rice, milled rice and parboiled rice is reported by several workers; Williams & Bevenue (1953) and Sone (1959). They reported that the amount of variation in dehusked rice is between 0.20 and 2.65 percent, 0.32 and 0.45 in polished rice and between 0.52 and 0.83 in parboiled rice.

5.3.4. Total free amino acid

Total free amino acids are minor but important chemical constituent mainly concentrated in the peripheral layers of the kernel of rice grain. Further milling to remove the pericarp, seed coat, testa, aleurone layer and embryo to yield white rice results in disproportionate loss of free amino acids and other components (Singh et al., 1998; Park et al., 2001). In the present investigation, free amino acid content differs significantly among the rice cultivars (Table III and Fig. 7). High content of total free amino acid is recorded in **Mpangban** {(T_1) 1.56 mg/g}, **De se** {(T_9) 1.56 mg/g} and minimum content is recorded in CAUR-3 {(T_{12}) 0.47 mg/g} and KD-2-6-3 {(T_{16}) 0.52 mg/g}. Low content of free amino acids in the grain may be because only few acids exist in free form and the others are bound as protein. Several workers like Basak et al. (2002) reported the protein content in the range of 6.6 to 7.3% in brown rice, from 6.2 to 6.9% in milled rice by Singh et al.(1998) and 8.2 to 8.4% in basmati rice by Deka et al. (2000).

5.3.4.0. Chemical Composition of *Albizia Bark*

Moisture content

Significant variation of moisture content is observed between *Albizia myriophylla* bark and *A. kalkora* bark

(Table IV Fig.8). Remarkably, *A. kalkora* bark has higher moisture content at 39.65% than *A. myriophylla* bark (27.10%). This is expected because the bark of *A. kalkora* is thicker than that of *A. myriophylla's*, which was found to be 1.5–3mm, and 1–2mm respectively. This could be the reason that more microorganisms are accumulated in *A. kalkora* bark. Similar observation was made by Pruthi (1969) that increase in the moisture content of grain favoured the development of insects as well as moulds. Pradhan (1969) also reported that moisture requirements for different insects varied from species to species.

Ash content

Significant variation of ash content in the two varieties of *Albizia* bark is recorded (Table IV and Fig.8). Significantly higher ash content is observed in *A. kalkora* bark (32.08 mg/g) than *A. myriophylla* (29.13%) possibly because *A. kalkora* bark may have higher mineral content than *A. myriophylla* bark. Hsu and Luh (1980) also reported the ash content of rice hull in the range of 13.2 - 29.0%.

Total free amino acid

A significant difference of total free amino acid content is observed (Table IV and Fig.9) in both the *Albizia* species. Higher content is recorded in *A. myriophylla* (0.11 mg/g) than in *A. kalkora* (0.07 mg/g). This could be because bark of the latter consists mainly of death cells.

Protein

Significant different protein contents are observed in the two types of *Albizia* bark powder (Table IV and Fig.9). *A. myriophylla* contains higher protein at 0.19 mg/g than *A. kalkora* at 0.12 mg/g. This could be because *A. kalkora* bark contains more cork cells which are death when matured.

5.4.1 Total phenol

Phenols, the aromatic compounds with hydroxyl groups

are widespread in plant kingdom. They are found in all parts of the plants. Phenolic compounds are important components of many fruits, vegetables, and beverage, to which they contribute to flavor, texture, color, and sensory properties such as bitterness and astringency (Floridi et al., 2003).

Significant variation of total phenol content is observed in the bark powder of two types of *Albizia* (Table IV and Fig.9). Higher content of total phenol is observed in *A. kalkora* (19.02 mg/g) than *A. myriophylla* (12.47 mg/g). The present results are comparable to the findings of Cai et al. (2003) who reported that total phenolic contents of 17 varieties of cow peas are in the range from 34.6 to 376.6 mg/ 100g of flour. Several phenolics have been isolated and identified from conifer needles by Lundgren (1987) and Strack et al. (1989), from bark of birch, spruce and pine by Pan (1995) and from white birch and silver birch leaves by Ossipov et al. (1996). Beyer et al. (1993) also reported a high isorhapontin concentration in the outer root bark, stem bark, and fine root extracts.

5.4.1.1 Acidic nature of Albizia bark

In the present investigation for acidic nature of the bark powder (Table V and Fig.10), p^H is interestingly found to be in the range of 4.8 to 5.03 in A. *myriophylla* and from 4.8 to 5.09 in *A. kalkora*. It is observed that increasing bark powder concentration from 1 to 10% does not make significant change in pH. The result indicates that compounds responsible for acidic nature of bark are weak acid. That could be the reason that mould and yeast have selected Albizia bark as their natural habitat.

5.4.1.2 Organic acids

Organic acids play a pivotal role in maintaining the quality and nutritional value of a variety of foods. They contribute to the sensory properties of foods. The organic acids present in foods originate from biochemical processes, their addition as acidulants, stabilizers or preservatives, or from the activity of some microorganisms.

From the present investigation for organic acid in the extracted sample of *Albizia* bark it is found that no organic acid could be detected. The reason could be that bark powder contains no organic acid or it may be present in a very very negligible quantity.

5.4.1.3 Phenolic acids

One of the most important phytochemical groups is phenolic. Plant phenolics are antioxidants in many food systems.

In the present investigation for phenolic acid from the bark powder extract, remarkably acids like caffeic, ferulic, gentisic, gallic, vanillic, salicylic, p-hydroxybenzoic, p-coumaric are detected from unhydrolysed and hydrolysed samples. Significantly, similar phenolic acids are found to be present in *A. myriophylla* and *A. kalkora* bark. Krygier et al. (1982) reported that free and esterified phenolics are soluble in methanol and found to be the predominant phenolics in some plants and hydrolysis will convert the esterified phenolics to simple phenolic acids which are in agreement with the present findings. Sahidi and Naczk (1995) also reported that plant phenolics are present in free, ester and insoluble bound forms. Rudkin and Nelson (1947) also isolated phenolics from sweet potato for the first time.

5.4.1.4 HPLC profile of phenolic acids

HPLC chromatograms of free phenolic acids in methanol extract before and after hydrolysis (Fig.11, 12 and Fig.13, 14) from bark powder are investigated. HPLC proves to be a valuable complement to TLC in identification or qualitative analysis of phenolic acids. Eleven phenolic acids are identified as gallic acid, gentisic acid, p-hydroxybenzoic acid, vanillic acid, caffeic acid, syringic acid, p-coumaric acid, ferulic acid, salicylic acid, querectin, eugenol and kaempferol. Other peaks are not identified. A large peak representing caffeic acid, p-hydroxybenzoic acid and vanillic acid are evident in *A. myriophylla* bark before hydrolysis. After hydrolysis large peak of vanillic acid, ferulic acid, synapic acid and gallic acid are observed. Compared with the chromatograms before hydrolysis showing only free acids, there is a large increase in the content of vanillic acid. This indicates that vanillic acid is the predominant phenolic acid in *A. myriophylla*, present mainly in esterified forms. A large peak representing salicylic acid is evident in *A. kalkora* bark representing both free salicylic acid and the one released after hydrolysis. A large peak of caffeic acid, vanillic acid and ferulic acid are also evident after hydrolysis, however, peak of caffeic acid and vanillic acid are found to overlap with each other. The result indicateds that salicylic acid is the most predominant phenolic acid in *A. kalkora* bark. Caffeic acid, vanillic acid and ferulic acid are also found to be predominant but in esterified forms. Chromatograms also indicate the retention times (RT) of all the phenolic acids. The elution of phenolic compounds follows the decreasing polarity in reversed-phase HPLC, so benzoic acid derivates are eluted earlier than cinnamic acid derivates. Similar findings are reported by Guo et al. (1997) that the retention time of phenolic compounds increases with the number of–OCH_3 substituents. Gallic acid is the first acid eluted (three - OH groups). Whereas vanillic acid, the first –OCH_3 substitutes benzoic acids and syringic acid (two –OCH_3 substituents).

How and Morr (1982) reported the presence of gentisic acid, syringic acid o–coumaric, salicylic acid, caffeic acid, p-coumaric acid and ferulic acid in soyabeans. Ribéreau - Gayon (1972); Harborne and Williams (1971); Glass and Bohm (1969) also reported that five derivatives of common hydroxybenzoic acid salicylic (2–OH), 4-hydroxybenzoic, vanillic (3-OMe, 4-OH), protocatechnic (3,4-diOH), and syringric acids (3,5-OMe, 4-OH) are widespread in the plant kingdom. Several workers like Medvedeva et al (1977); Neimann (1969), Gromova et al. (1977) and Ivanova et al. (1965) reported the presence of glucosides of 4-hydroxybenzoic, vanillic and protocatechuic acids in needles of conifers (species of Larix, Abies and Pinus).

5.5.1 Total microbial load

In the present investigation higher total microbial loads (Table VI, Fig.15) in *A. kalkora* bark powder {9 x 10^3 cfu/g (LB), 1.88 x 10^4 cfu/g (YPDA)} than *A. myriophylla* bark powder {80 cfu/g (LB), 1.2 x 10^2 cfu/g (YPDA)} are observed. Since *A. kalkora* has higher moisture and ash content, it might have provided a better environment for mould and yeast to survive. The result indicates that fungi are more dominating than bacteria in the bark; this may be because of the p^H of the bark, which ranges from 4 to 5 is an optimum range for the growth of fungi. The present results are in agreement with the findings of Qi et al. (2002) who reported that maximum growth for most fungi was observed at a p^H of approximately 5.0.

5.5.1.1 Isolation and characterization of microorganisms

In the present investigation altogether five putative yeasts or mould strains are isolated from the bark powder of both *Albizia spp.* (Table VII). These isolates were identified at the Institute of Microbial Technology (IMTECH), Chandigarh (India). They are *Mucor indicus* Lendner (Mold) which has colourless mycelium, fast growing and sporulating, isolated from *A. kalkora* bark; *Candida parapsilosis* which is medium growing with white colony, *Rhodotorula mucilaginosa:* also medium growing with pink colony, and *Streptococcus sp:* colourless and slow growing. The present results are in agreement with the work of Sanchez (1977) & Tanimura et al. (1978b) who found a mixture of yeasts, bacteria and mould in *M. gradifola* Linn leaves, bark or fruit used on Phillipine **Basi**. Sie (1962) and Soedarsono (1972) also reported that raw dry spices contain a wide variety of microorganisms and the species, when added to some **ragis** contribute other microorganism or may inhibit development of undesirable microorganisms. Went and Prinsen Geerligs (1895); Ellis et al. (1976) and Wehmer (1900 a.b) also isolated *Amylomyces rouxii, Endomycopsis fibuliger, E. burtonii Candida parapsilosis, C. melinii, C. solani Mucor rouxii, Hansenula sp., Rhizopus sp.* from Indonesian **ragi**. Dahiya and Prabhu (1977) reported the presence of *Rhizopus sp.* and *Mucor sp.* in Indian **Madhu** and Indian rice **Bur-Pachwai**. Uyenco and Gacutan (1977b) isolated *Rhodotorula glutinis, Candida parapsilosis* from Phillipine **Tapuy**.

5.5.1.2 Fermentation of individual carbohydrates by isolated microorganisms

In the present investigation different fermentation abilities are significantly observed among the isolated microorganisms (Table VIII). From the observation it is remarkably indicated that *M. indicus* produce maximum bubbles in the tube of glucose, glactose, maltose and sucrose but not in lactose. *R. Mucilaginosa* are also found to produce bubble in glucose, galactose and lactose but not in maltose and sucrose. *C. Parapsilosis* are found to produce bubbles only in glucose and galactose. Another remarkable finding is that *Streptococcus sp* could produce bubble only in lactose. In all cases, maximum bubble formation is observed upto 120 hours of incubation after which, it decreases. Probably, mould (*M. indicus*) possesses both amylolytic and saccharolytic activities which is why they can degrade starch and ferment it to alcohol. Similar findings are also reported by Saono (1977) that the entire mould isolates from Indonesia **ragi** and **tapé** are amylolytic, lipolytic and proteolytic. Yeasts are known to have fermentative ability but only few are amylolytic and lipolytic. This could be the reason why *C. parapsilosis and R. mucilaginosa* are not able to produce bubble in maltose and sucrose. It is found that *Streptococcus sp.* could ferment only lactose but not the other carbohydrates. The findings are in conformity with the finding of Kinsella and Hwang (1976) who reported that *Streptoccoccus lactis* degrade the lactose to produce mostly lactic acid, some citric acid and perhaps some ethanol. Decrease of bubble formation after 120 hours of incubation time could be due to lack of substrate.

5.5.1.3 Hamei and 'Yu' (Zouziu) fermentation

The flow sheet for laboratory preparation of ***hamei*** and ***'Yu'*** (*Zouziu*) is presented in Fig.4. The procedure is the result of extensive preliminary test designed for efficient fermentation and to optimise the yield. In the process, raw material (rice and bark) is crushed in mechanical grinder into powder and gamma sterilised. Pure culture of microorganisms *M. indicus* is used in preparing ***hamei*** (starter) culture. Radiation sterilisation of bark and rice powders is introduced to get rid of unwanted microorganisms. However, cooking of rice would take care of the contamination in rice. Mould-inoculated ***hamei*** is inoculated and mixed into cooked rice. The incubated mixture (for 5 days at 26 ± 2°C) is harvested and the rice beer thus obtained is distilled at 78°C to produce the final product ***"Yu"***. This method eliminates the tedious method of traditional fermentation. Traditional methods are conducted under unsterilised condition and different organisms acted in concert resulting in low yield and quality. Hesseltine et al. (1963), Steinkraus et al. (1960) and Wai (1964,1929) developed pure culture fermentation of **tempeh** and **sufu.** Ko (1977) also produced **ragi** containing pure cultures of single microorganisms. Wang et al. (1975a) developed a method for production of *R. oligosporus* spores for the **tempe** fermentation. Because of problems faced in controlling the natural fermentation, pure culture techniques have been developed in recent years. Kang et al. (1999) tried to modify maize starch by ã-irradiation and p^H adjustment and found that the combination considerably affected the viscosity. Bao et al. (2001); El-Saadany et al. (1976); Bachman et al. (1997) and Sokhey et al. (1993) reported that increased radiation dose is invariably associated with lower viscosity and higher solubility and acidity of starches. Sanchez-Marroquin (1953), reported that optimum temperature was found to be 28°C (82°F) and optimum P^H at 5.0 under pure culture conditions using pasteurised **Agave** juice; fermentation time was 48 to 72 hours versus 8 to 30 days for natural fermentations.

5.5.1.4 Fermentation of rice by isolated microorganisms

In the present investigation of the final product of rice fermentation by isolated microorganisms (Fig.16), the chromatograms of headspace volatile compounds (ethanol) of the distillate analysed by GCMS show that the peak obtained from fermentation of *M. indicus* inoculated ***hamei*** is found to match with the ethanol peak available in the GCMS library (Fig.16,17). Fermentation using yeast and bacteria (*C. parapsitosis* and *R. mucilaginosa*) does not give any resembling peak (Fig.18,19, 20, 21) but when mould is mixed with yeast and bacteria in the fermentation, the resembling peak is obtained again. (Fig.22, 23, 24, 25, 26). The results indicate that mould (*M. indicus*) is found to be the key factor for the alcoholic fermentation. Probably they possess both amylolytic and saccharolytic activities to saccharify cooked rice and rice powder. The present findings are in conformity with the findings of Auerman (1972) that amylolytic enzymes are responsible for hydrolysis of starch to sugars and these degraded starches are converted into ethanol. Yeasts are known to perform fermentation in the presence of glucose but lacks in amylolytic activities, hence unable to carry out alcoholic fermentation by themselves.

5.5.1.3 Quantification of ethanol in distill product by enzymatic assay

The enzymatic assay of the fermented product of isolated microorganisms (Table IX Fig.27) shows that there are significant variations of ethanol yield. It is observed that in fermentation, *M. indicus* alone produce 65.43%. Fermentation of rice by *Streptococcus sp.* produces 0.14%, *C. parapsilosis* produce 0.34% and *R. mucilaginosa* produce 1.60%. Significantly high yield of ethanol is observed in the combination of mould and yeast: *M. indicus* + *C. parapsilosis* produce 65.51%, *M. inducus* + *R. mucilaginosa* produce 67.22% and *M. inducus* + *Streptococcus sp.* produce 62.49%. The present findings are in agreement with the finding of Yeoh (1977b), Saono et al. (1977), who reported the effect of various yeasts in combination with *Amylomyces rouxii* on ethanol production on rice at 30°C. On critical analysis of the findings it is found that *M. indicus* alone produce equally similar percentage of ethanol.

5.5.2.0 Factors Affecting Ethanol Yield

5.5.2.1 Albizia variety, water quantity and mode of inoculum in fermentation

In the present investigation (Table X and Fig. 28) it is observed that there are no significant differences of ethanol yield from two types of bark (*A. myriophylla* and *A. kalkora*). Ethanol yield from ***hamei*** of *A. myriophylla* bark is 75.62% and from that of *A. kalkora* is 76.51%. Direct adding of bark powder in the broth along with 1 ml mould suspension inoculum also gives more or less similar yield i.e *A. kalkora* produces 70.98% and *A. myriophylla* produces 70.18%. The reason may be that bark powder of the two species possess more or less similar biochemical properties, which are helpful for the growth and metabolism of the organisms.

Water volume in fermentation broth has a significant effect on ethanol yield. 1:5 w/v of water is observed as optimum for optimum yield than 1:10 w/v. This could be due to abundant sporulation of mould at this moisture level. Similar findings are made by Wang et al. (1975) that *R. oligosporus* sporulate abundantly on rice, wheat bran and soyabeans at a moisture level of 40% as well as 50%, but not at 30% and the moisture level in medium also determines the vegetative growth and sporulation of organisms.

Huang et al. (1985) also reported that even in solid-state fermentation, water must be available for microbial activities and increase in moisture level from 50 to 55% has resulted in 1.64-fold increase in gluco-amylase production. Lousane et al. (1985) also found that moisture requirement is governed by nature of the substrate type of end product and the microorganism. Several workers like Jansz et al. (1977); Ramadas et al. (1996); Kolhekar et al. (1985); Oriol et al. (1988); Ghildyal et al. (1985) also reported that glucoamylase is produced by Solid State Fermentation (SSF) process. Inoculum in the form of ***hamei*** is found to be more effective than 1 ml mould suspension. The reason could be that ***hamei*** has higher total microbial load than 1 ml suspension. The present findings are in conformity with the finding of Arasaratnam et al. (2001) that the number of spores inoculated should be optimised, and having higher or lower levels of spores may affect and delay the enzyme production. Martinelli and Hesseltine (1964) also reported that at the inoculum level of 1×10^6 spores per 100 gm cooked soyabeans, good **tampeh** is made by tray and package fermentaton.

When critically observed it is found that adding bark powder along with 1 ml suspension inoculum gives higher yield at 70.98% and 70.18% for *A. kalkora* and *A. myriophylla* than without bark powder at 60.29% and 54.98% respectively. This could be due to the acidic nature of bark powder which will contribute to a slight variaton of p^H in the broth and provide an environment for mould to produce enzyme. Similar findings are reported by Frank et al. (1976)

that optimum pH for enzyme is 6.5 to 7.0; and Hoover (2000) also reported that depolymerisation of starches is higher in the presence of acid. Devi (1980) also reported that during fermentation of ***hamei*** amylase enzyme at pH 3.5, 5 and 6.5 was active and plays the role in the accumulation of glucose. Amylase enzymes having optima at PH 5.5 and 6.5 in *Lactobacillus* are reported by Champ et al. (1983).

5.5.2.2 Bark powder quantity in hamei

Significant variation of ethanol yield with respect to bark powder percentage in ***hamei*** is observed (Table XI and Fig.29). In the present investigation, remarkably highest ethanol yields of 76.98% is obtained from 2% bark powder concentration and minimum yield of 69.52% from 5% concentration, while 1% and 3% concentration gives a comparatively equal yield of 73.38% and 73.95% respectively. The reason could be that bark powder containing phenolic compounds of higher concentration may be toxic for the microorganisms. The findings are in conformity with the finding of Alexander (1985) that phenol in sufficiently high concentrations can be a toxicant for species capable of using it as growth substratum.

A number of microorganisms have also been reported to degrade phenol at low concentrations including *Alcaligenous eutrophus* by Hughes et al. (1984), *Baccillus Stearothermophilus* by Burwell (1975), *Pseudomonas sp* by Bayly and Wigmore (1973), Bettman and Rehm, (1984); Feist and Hageman (1969), in *Rhodococcus sp* by Straube (1987) and *Trichosporan cutaneum* by Neujahr and Varga (1970), Anderson and Dagley (1980); Cain (1980); and Jones et al. (1993). Middlehoven (1993) also reported the degradation or biotransformation of phenolic pollutants in aquatic and terrestrial environments by mycelial fungi and yeast. The present results are in agreement with many investigators: Kasenberg and Traguair (1987) who reported the growth inhibition of fungi by caffeic acid and Bowles and Miller (1994), Reinders et al. (2001) reported for bacteria.

5.5.2.3 Incubation period

Analysis on ethanol yield with respect to incubation period of ***hamei*** and fermentation broth (Table XII and Fig. 30) reveals a significant observation. From the finding it is indicated that maximum ethanol yield is obtained from 120 hours incubation period; after this, yield starts decreasing. The reason could be due to decrease of enzymes activities and increase of alcohol concentration which inhibits viability of cells and thereby affecting the yield. The findings are in conformity with the findings of Wang and Vespa (1974) that enzyme production of *R. oligosporus* grown on wheat bran at 50% moisture and 25°C were at maximum for 3 to 4 days, *M. dispersus* for 3 to 4 days and *A. elegans* for 3 days. Cronk et al. (1977, 1979) also reported that mould itself was able to produce an alcohol content of 5.6% v/v at 96 hours.

5.5.2.4 Rice quality and quantity of hamei inoculum

Quality of rice in ***hamei*** and fermentation broth and quantity of ***hamei*** inoculum has a significant effect on ethanol yield (Table XIII and Fig. 31). Higher yields are obtained from 12 gm ***hamei*** and fermentation broth of unpolished rice (75.68%) than 12 gm ***hamei*** and fermentation broth of polished rice (66.71%). When crosschecked for 12 gm unpolished rice ***hamei*** and polished rice broth it gives 70.09%, and 71.74% in vice versa. When ***hamei*** is reduced to a 6 gm, it is observed that 6 gm ***hamei*** of unpolished rice and broth still give higher yield of (73.76%) than 6 gm ***hamei*** of polished rice and broth (67.64%). When it is crosschecked again for 6 gm unpolished rice ***hamei*** and polished rice broth, it gives 69.79%, and 70.33% in vice versa. From the observation it is found that unpolished rice (local variety) and 12 gm of ***hamei*** (unpolished and polished) give better ethanol percentage. This is expected because, with rice, the bran in the outer layer of grain contains relatively large amount of protein, fat, ash, vitamins and minerals which may serve as nutrient for the growth and metabolism of microorganisms. Milling and polishing remove the bran from the grain. A lower degree of milled rice has a greater nutritional value.

A mature ***hamei*** contains approximately 6.2×10^6 cfu/g of *M. indicus* which seems to be ideal for maximum production of enzymes for starch degradation, and this may be the reason that 12 gm of ***hamei*** give better yield of ethanol than 6 gm ***hamei.*** The finding can be compared with the findings of Oriol et al. (1988); Huang et al. (1985); Gervais and Sarrette (1990); Shanmuganathan et al. (1994) and Tao et al. (1997), that when the number of spore is increased from 4.5×10^7 to 4.5×10^8 ml^{-1}, maximum enzymes production is obtained on the 2nd day, while with spore of 4.5×10^5 and 4.5×10^9 ml^{-1}, maximum activity is obtained on the 3rd day. Park et al. (2002) also reported that an inoculum size of 5×10^5 spores/gram, initial moisture content of 65%, cultivation time of 5 days and 10 times concentration of basal medium are optimum for Xylanase production in solid-state fermentation.

5.5.2.5 Rice variety in Hamei and fermentation broth

In comparative study of different rice cultivars in ***hamei*** and fermentation broth for ethanol yield (Table XIV and Fig.32), maximum yields are observed in ***hamei*** from the rice varieties of RCM-9 {(T_{15}) 75.29%}, KD-2-6-3 {(T_{10}) 74.62%} and **De se** {(T_9) 73.54%} and minimum in IR-64 at {(T_{13}) 65.55%}. In the fermentation broth, maximum yields are recorded in RCM-5 {(T_{14}) 76.22%} than in RCM-9 {(T_{15}) 75.58%}, RCM-10 {(T_{16}) 75.53%}, KD-2-6-3 {(T_{10}) 74.21%} and **De se** {(T_9) 74.55%} and minimum is recorded in **Ritiang** {(T_5) 68.38%}.

Present data indicate that rice cultivars / varieties of RCM group and KD-2-6-3 from the lowland rice and **De se** from upland rice, which are semi glutinous, are the best substrates for ***hamei*** and rice fermentation. Other cultivars like **Mpangban** (T_1), **Tunanapnang** (T_2), **Napdai** (T_3), which are waxy and highly glutinous give a low yield. Since high concentrations of sugar and ethanol in the broth are not favourable for fermentation, this could be the reason that these rice varieties are not favourable for fermentation. The findings are in conformity with the finding of Atiyeh and Duvnjak (2001) that specific growth rate and biomass yield of yeast decreases when sucrose concentration increases due to growth inhibition by a high substrate concentration. Wong and Jackson (1977a), Pichyangkura and Kulprecha (1977) also reported that Philippines **bubud** is made either from glutinous rice mixtures of glutinous and ordinary rice or from rice flour.

Kodoma and Yoshizawa (1977) also reported that rice of short-grain varieties is considered best for **sake** manufacture and large grains are considered desirable. Kadoma (1966a) revealed that desirable properties of rice for brewing **sake** are the soft kernel (which the *mycelia* of the **koji** mould can easily penetrate), high rate and large percentage of water absorption. These findings are comparable with the present finding that KD-2-6-3 (T_{10}) which has soft kernel serves as the best rice for making ***hamei*** and RCM-5 (T_{14}) RCM-9 (T_{15}) RCM-10 (T_{16}) which are medium-size grains with soft kernel, and **Dese,** a short-grain variety produce the best yield in fermentation broth.

5.5.2.6 Hamei of Phenolic acid

Hamei made of phenolic acids substituting *Albizia* bark has a significant effect on ethanol yield (TableXV and Fig.33). It is observed that individual phenolic acid (2% caffeid acid, 2% ferulic acid, 2% vanillic acid and 2% salicylic acid) per ***hamei*** gives lesser yield (53.86%, 51.17%, 52.85% and 53.02%) respectively than ***hamei*** of acid concoction (76.32%).

Remarkably it gives a better yield than bark powder. The present finding indicates that bark powder can be substituted by phenolic acids in rice fermentation, thus, supplementing the fermenting medium and create a favourable environment for maximum production of enzymes and starch saccharification. Similar findings are also reported by Hoover (2000) that depolymerisation of starches is higher in the presence of acid. Acids such as HCl and H_2SO_4 cause scission of the glucosidic linkages thereby altering the structure and properties of the native starch. Azaharand Handy (1981) also used acid treatment for starch hydrolysis. Steeping in acid partially degraded the starch particles making them susceptible to the action of enzymes.

5.6.0 Morphological and Physiological features of Mucor indicus

Colony morphology of *M.indicus* is shown in Plate 30. The colony is a colourless, hairy mycelia which displays the character of typical mould sp. The diameter of the colonies grown overnight is found to vary from 1.4 to 2.5 cm, but it increases significantly on further incubation.

5.6.1 Growth kinetics of M. indicus in hamei and fermentation broth with respect to inocubation time

Growth kinetics of *M.indicus* in ***hamei*** and fermentation broth with respect to incubation time is studied (Table XVI and Fig. 34). Colony forming unit (cfu) of *M.indicus* in ***hamei*** with bark powder ($2.95x10^6$ cfu/g) is found to be higher than in ***hamei*** without bark powder ($1.2x10^6$ cfu/g) at the end of 24-hour incubation. After 96 hours it is remarkably increased to $6.2x10^6$ cfu/g and $5.7x10^6$ cfu/g respectively. However, after 120 hours of incubation it is found to decrease to $5.3x10^6$ cfu/g and $4.45x10^6$ cfu/g. The finding is in agreement with the finding of Kim (1968) that yeast increases rapidly to $2.0x10^8$ cells/ml after 24 hours and $8.4x10^8$ cells/ml in 48 hours. Yeasts numbers decrease slightly to 5 to $7x10^8$ cells/ml at the end of fermentation. Lactic acid bacteria increase after 24 hours, reaching $2.0x10^8$ cells/ml by 3rd or 4th day but decrease thereafter to $4.0x10^5$ cells/ml. Merican (1977) also reported that yeast count of fresh **toddy** ranges from $1.0x10^8$ to $3.1x10^9$ and bacteria ranges from $1.5x10^8$ to $3.6x10^9$ per ml.

Initial Cfu counts in fermentation broth inoculated by ***hamei*** with bark powder and ***hamei*** without bark powder after 24 hours is $2.3x10^6$ cfu/ml and $2.2x10^6$ cfu/ml; it increases to $6.6x10^6$ cfu/ml and $5.8x10^6$ cfu/ml at 96 hours incubation and decreases to $6.3x10^6$ cfu/ml and $5.3x10^6$ cfu/ml respectively. The reason may be that at the end of fermentation, ethanol content in the broth reaches upto 18 to 20%, which is toxic for the organisms to survive. In the investigation it is also found that significantly higher cfu counts are observed from ***hamei*** made of bark powder and fermentation broth inoculated by this ***hamei*** at 96 hours incubation. Lee et al. (1969) reported that in **yakju (takju)** fermentation, 70 to 80% fermentation is completed within 48 hours and 90% complete within 72 to 96 hours. Hong et al. (1968a) also reported that most of the starch is converted to sugar within 12 to 24 hours. Ethanol continues to increase through out the fermentation.

5.6.2 Temperature tolerence of M.indicus

Temperature tolerance of *M. indicus* as observed in Table XVII indicates that optimum growth temperature of *M .indicus* is found in the range of 26 to 35°C. At a temperature below 10°C the organism is not able to grow, but does not lose its viability and at 40°C and above, organism completely lost its viability. Similar findings are also reported by Hesseltine et al. (1974) that *M. indicus* grow fairly well on wheat bran of 35% moisture at 25 to 32°C, but grow slower at 20°C; they produce impressive amount of enzymes at 20 or 25°C but significantly less at 32°C. Enzyme activities of *R. oligosporus* reach a maximum after 2 to 3 days at 32°C, 3 to

4 days at 25°C and 5 to 7 days at 20°C, but they yield the highest amount of protease when grown on wheat bran of 50% moisture for 3 to 4 days at 25°C.

5.6.2 pH tolerance of M.indicus

In the present investigation (Table XVIII) it is found that *M.indicus* grow well between pH ranges of 5 to 9. At pH 10 the growth is slow while between 3 to 4 pH some growths are observed when incubated for 48 and 72 hours. Even at pH 2, viability is observed but not growth. The optimum pH range for these organisms is between 5 to 9 and this may the reason that they thrive in *Albizia* bark as their natural habitat or natural substrate whose pH is between 4.5 –5.

5.6.4 Ethanol tolerance of M. indicus

Since *M.indicus* is found to produce high yield of ethanol, its tolerance limit is investigated (Table XIX and Fig.35, 36). Ethanol tolerance is determined by treating with different concentration of ethanol for different period of time. The organism is found to be tolerant upto 25% concentration at 2 hours treatment. At 15% concentration it is found to be completely tolerant up to 24 hours treatment but decrease in colony count (cfu) is observed as the treatment duration increases. At 20% concentration it is found to be tolerant upto 6 hours but beyond this period it is lethal. At 25% concentration, a few viable colonies are observed at 2 hours treatment; beyond this period it results in a complete loss of cell viability. The reason could be that at 20% concentration and above, it inhibits cell growth and cell viability. The present findings are in support with the finding of Ingram and Buttke (1984) and Osman and Ingram (1985) that ethanol causes the leakages of essential cofactors and coenzymes which are essential for the activity of enzymes involved in glucolysis and alcohol production on fermentation in *Zymomonas Mobilis* as well as in yeast.

6
Summary and Conclusion

Alcoholic derived from fermentation received a worldwide attention as an efficient alternative energy source, as a laboratory solvent and as an antiseptic since several decades. Attempts to increase ethanol concentration during fermentation adversely affects growth, viability and fermentation rate of (yeast) cells. In the fermentation of **'Yu'**, raw materials (*Albizia* bark and rice variety) have a significant effect on ethanol production. The works done on the topic "**Study on the role of *Albizia spp.* in the fermentation process of different of rice cultivars**" have been summarized below:

During the present investigation various techniques were employed for various biochemical studies of the raw materials and estimation of the final product. *Albizia* plant was identified according to Kanjilal and Das (1939); percentage of moisture and ash was estimated as per AOAC (1970) method, while starch estimation was done by the method of Chopra and Konwar (1976), and amylose was estimated by using that of Sadasivam & Manickam (1992) and Swambhagya & Bhattacharya (1979). Total soluble sugar was estimated by Dubois's et al. method (1951), reducing sugar by Somogyi's (1952) and Nelson's method (1944), non reducing sugar by Malhotra's and Sarkar's method (1979), total free amino acid by Yemn and Cocking's method (1955), Protein by Lowry's et al. method (1951), total phenol by Bray and Thorpe's method (1954), phenolic acid by Variyar et al. method (1988), organic acids by Ryan method (1972), Head space volatile by Kang et al. modified method (2003), and enzymatic assay by Bernt and Gutmann's method (1974).

1. Two species of *Albizia* plant commonly used by brewers in rice fermentation were collected and identified as *A. myriophylla* and *A. kalkora*. 16 (sixteen) rice cultivars commonly used were also collected and out of these, nine were from hills (upland rice) and seven were from the valley (plain).

2. Biochemical analysis of rice shows significant variation of chemical composition among the grains. Maximum moisture percentage was recorded in **Mpangban** {(T_1) 11.59%} and minimum was recorded in **Ritiang** {(T_5) 7.4%}. The highest ash content was observed in KD-2-6-3 {(T_{10}) 1.96%} and lowest in **De se** {(T_9) 0.96%}; highest starch content was observed in **Mpangban** {(T_1) 88.48mg/g} while the lowest content was found in IR-64 {(T_{13}) 69.89mg/g}. Amylose content was found to be highest in IR-64 {(T_{13}) 26.71mg/g} and lowest in **Mpangban** {(T_1) 14.28mg/g}. Maximum reducing sugar was recorded in **Mpangban** {(T_1) 0.52mg/g} and minimum was recorded in IR-64 {(T_{13}) 0.14mg/g}. RCM-9 (T_{15}) was found to contain maximum non reducing sugar at 0.89mg/g and **Sangbu** (T_8) was found to contain the lowest at 0.47mg/g. Highest content of total free amino acid was recorded in **Mpangban** {(T_1) 1.56mg/g} and CAUR-3 has the lowest at 0.47mg/g.

3. Biochemical analysis of *Albizia* bark powder also enumerated the variation of chemical composition between the two species. *A. kalkora* bark was found to

contain higher moisture percentage {39.65}, ash percentage (32.08), and total phenol (19.02mg/g) than *A. myriophylla* which contains 27.10% moisture, 29.13% ash and 12.47mg/g of total phenol. However, higher contents of total free amino acid and protein were recorded in *A.myriophylla* (0.23mg/g and 0.19mg/g) than *A. kalkora* (0.07mg/g and 0.02mg/g).

4. Since bark powder is found to have acidic nature, investigation for acid (Organic acid and Phenolic acid) was done. On analysis, organic acid was not dectected in the bark but different phenolic acids were detected. Caffiec acid, gallic acid, ferulic acid, gentisic acid, p-hydroxybenzoic acid, syringic acid, p-coumaric acid, vanillic acid, salicylic acid, quercetin, eugenol and kaempferol were found to present in bound and free forms.

5. Isolation and Identification of microorganisms: One mould, three yeasts and one bacterium were isolated from *Albizia* bark powder and ***hamei***. They are *Mucor indicus* Lendner, *Candida parapsilosis, Rhodotorula mucilaginosa, R. mucilaginosa* and *Streptococcus sp.* These isolates were identified from Institute of Microbial Technology (IMTECH), Chandigarh, India.

6. The isolated microorganisms were screened for their fermentation ability on individual carbohydrates- glucose, galactose, lactose, maltose and sucrose. Maximum fermentation was observed in mould and minimum was observed in bacteria. Except lactose, mould could ferment glucose galactose, maltose and sucrose, *Candida parapsilosis* ferment glucose and galactose, *Rhodotorula mucilaginosa* ferment glucose, galactose and lactose, and *Streptococcus sp.* could ferment only lactose.

7. Fermentation abilities of rice were also tried for the isolates. For this purpose ***Hamei*** was prepared by

inoculating these isolates and subsequently inoculated in cooked rice. On analysis of the fermented product it was found that only mould could produce ethanol. Yeast and bacteria were not able to ferment rice by itself. But when mould was combined with other isolates (yeast and bacteria) it was found to produce ethanol. Qualitative estimation of ethanol in the final product was done by GCMS Head Space method, and quantified by enzymatic assay (alcohol dehydrogenase) method. Mould alone was found to produce about 65.43% of ethanol but other isolates produced very negligible percentage. If mould was combined with other isolates it produced ethanol in the range of 62.49% and 67.22%.

8. Factors affecting ethanol yield: Since mould alone was found to carry out fermentation efficiently and successfully, pure culture of mould was used for all the fermentation experiment.

Bark powder of two types of *Albizia sp.* (*A. myriophylla* and *A. kalkora*) in the fermentation produced more or less similar ethanol percentage at 75.62% and 76.51% respectively. Inoculums in the form of ***hamei*** were found to be more effective than 1 ml suspension inoculum, which produced 76.51% and 60.29% respectively. Remarkably, increasing yield of ethanol was observed when 1 ml inoculum was supplemented by bark powder in the broth which increase to 70.18%.

Water ratios at the time of cooking rice also significantly affect the yield. In all the conditions 1:5 w/v water ratio gave the best yield of ethanol.

On investigation for bark powder concentration with respect to ethanol yield, it was found that 2% gave the highest yield of 76.89%, but the yield was found to decrease at 5% concentration (69.52%). Periodic determination of ethanol yield from ***hamei*** and fermentation broth from day 1 to day

5 showed that ethanol yield increased with the increase of incubation period. Maximum yield was observed after 120 hours of incubation which was 77.35% in case of ***hamei*** and 76.27% in case of broth, but after this period the yield decreased to 73.61% and 72.33% respectively.

Quality of rice and quantity of ***hamei*** inoculum were found to be very important for fermentation. Local variety (unpolished) gives higher percentage of ethanol 75.68% than the milled market rice (66.71%). ***Hamei*** and cooked rice at a ratio of 1:4 produced higher yield of ethanol (75.68%) than at the ratio of 1:8 ratio {(***Hamei***: cooked rice) 73.76%}.

16 (sixteen) rice cultivars were screened for the best raw material in the fermentation with respect to ethanol yield. Variety RCM-9, KD-2-6-3, and **De se** were found to be the best for making ***hamei*** and variety RCM-5, RCM-9, RCM-10 and KD-2-6-3 were found to be the best for rice fermentation.

Since phenolic acids were found to be present in the bark powder, ***hamei*** was prepared by adding artificial phenolic acids and used for rice fermentation. The study showed that ***hamei*** made of phenolic acids produced almost the same amount of ethanol as that of ***hamei*** made from *Albizia* bark powder, i.e. 76.32% for the former and 75.09% for the latter.

9. Morphological and physiological feature of *M.indicus: M.indicus* shows typical characters of mould sp; they are a colourless, hairy mycelia with fast sporulation. The diameter of the colonies grown overnight varies from 1.4 to 2.5cm, which increased significantly on further incubation.

Growth kinetics of *M. indicus* in ***hamei*** and fermentation broth with respect to incubation time was studied. Maximum growth and sporulation were observed at 96 hours. Colony counts cfu were found to be 6.2×10^6 cfu/g in ***hamei*** and 5.7×10^6cfu/ml in broth. When incubation period was increased to 120 hours, colony count decreased to $5.3x10^6$cfu/g and $4.45x10^6$ cfu/g respectively.

Optimum temperature for growth and sporulation of *M. indicus* was found to be between 26 and 35°C. Organism was not able to grow at a temperature below 10°C and its viability was lost at the temperature of above 40°C.

M. indicus were found to tolerate pH from of 2 to 10. But optimum range was from 5 to 9. At pH 10, 3 and 4 some growths were observed. Since *M. indicus* can produce high percentage of ethanol, their tolerance to ethanol was investigated. It was found that they could tolerate upto 15% of it. On further investigation it was found to tolerate upto 25% but for a very short period of time (2 hours) and tolerable upto 9 hours at 20%.

Rice wine brewing in Manipur is a natural fermentation which have been handed down from generation to generation. It has been carried out largely on the basis of traditional techniques and there is no scientific approach. Therefore, there is a need to introduce modern technology to improve the yield. From the above investigation it may be concluded that *Albizia* bark powder is an indispensable raw material for making of ***hamei***. Bark of *A. kalkora* and *A. myriophylla* possess similar chemical composition and provide microorganisms for fermentation. Phenolic acid maintains acidic nature of the bark and provides a favourable environment for the fungus, yeast and mould to select as their habitat. Fermentation of rice is carried out by *M. indicus* alone which is the sole organisms for ***hamei*** and cooked rice fermentation. Short-grain, medium-grain, semi-glutinous varieties of rice and unpolished ones (low milling percentage) are favourable for brewing. For improving yield, *Albizia* bark can be substituted by artificial phenolic acids in ***hamei*** preparation; inoculum in the form of ***hamei*** culture, maintenance of temperature at 22±2 to 26± 2°C, and pH at the range of 5 to 6.5 using 1:5 water volume in cooking and incubation of 5 days are found to be ideal.

Recent knowledge of the morphology, physiology, biochemistry and ecology of *M. indicus* enable us to develop the conventional method of rice brewing to some extent. This scientific approach may also make further improvements possible in the technology of **'Yu'** brewing by eliminating certain irrational procedures and bring about revolutionary changes in the technology which will result in the domestic breweries being replaced by large-scale enterprise.

Bibliography

AOAC. (1965). *Official Method of Analysis* (9th Ed.) Association of Official Analytical Chemists, Washington D. C.

AOAC. (1970). *Official Method of Analysis* (10th Ed.) Association of Official Analytical Chemists, Washington D. C.

AACC. (1962). *Approved methods of the AACC.* (7th Ed.) American Association of Cereal Chemists, St. Paul, MN.

Adegoke, G. O. Nwaigwe, R. N. and Oguntimein, G. B. (1995). Microbiological and biochemical changes during the production of sekete –a fermented beverage made from maize. *Journal of Food Science and Technology*, India, 32 : 516 –518.

Aidoo, K. E. ; Hendry, R. and Wood, B. J. B. (1982). Solid substrate fermentations Adv.. *Appl. Microbiol.* 200: 230-237.

Akingbala, J. O. ; Rooney, L. W. and Faubion, J. M. (1981). A Laboratory procedure for the preparation of Ogi, A Nigerian Fermented Food. *Journal of Food Science.* 46: 1523 – 1526.

Akinrele, I. A. (1970). Fermentation studies on maize during the preparation of a traditional African Starch Cake food. *J. Sci. Food Agri.* 21 : 619.

Al- Bayati, B. and Al-Rayees, H. (1981). The chemical composition of Iraqi rice and rice by-products. *J. Food Sci. Tech.* 18 (2): 40-44.

Alexander, M. (1985). Biodegradation of organic chemicals. *Environ. Sci. Technol.,* 8, 106-111.

Alexandre, H. and Rousseauxl, Charpenties, C. (1994). Relationship between ethanol tolerance, lipid composition and plasma membrane fluidity in *S. cerevisiae* and *K. apiculata. FEMS Microbiol Lett* 124: 17-22.

American Association of Cereal Chemist. (2000). *Approved Methods of the AACC,* 10th ed.; Method 61-02. The Association: St. Paul, MN,

Amerine , M. A. ; Berg, H. W. ; Kunkee, R. E. ; Ough, C. S.; Singleton, V. L. and Webb, A. D. (1982) *The technology of Wine Making,* 4th ed. AvI Publishing Company, Inc. , Wesport, Conn.

Amerine, M.A. and Ough, C. S. (1980). Phenolic compounds. *In methods of Analysis Must and Wines.* Wiley: New York, Chapter 7, pp 175-199.

Anderson, J.J. and Dagley, S. (1980). Catobolism of aromatic acids in *Trichosporm cutaneum. J. Bacteriol.* 141: 534-543.

Anderson, R.A. (1976). Wild rice: Nutritional review. *Cereal chem.* 53:949-955.

Andrade, P. B. ; Seabra, R. M. ; Valentão, P. and Areias, F. (1998). Simultenous determination of flavonoids, Phenolic acids and coumaric in sevenmedicinal species by HPLC/diode array detector. *J. Liq. Chrom. Relat. Technol.* 21, 2813-2880

AOAC. (1975). *Official Methods of Analysis,* (12th Edn). Association of Official Analytical Chemists, Washington, D. C.

Arasaratnam, V. (1989). Bioconversion of Starch to Glucose and glucose as feed stock for the production of ethanol, lactic acid and fructose. Ph.D. Thesis, University of Jaffna, Sri Lanka.

Arasaratnam, V. ; Mylvaganam, K. and Balasubramaniam, K. (2001). Glucoamylase production by *Aspergillus niger* in solid state fermentation with paddy husk as support. *J. Food Sci Technol.,* 38. (4): 334-338.

Arasaratnam, V., Mylvaganam, K. and Balasubramaniam, K. (1997). Paddy husk support and rice brand for the production of glucoamylase by *Aspergillus niger*. Internatl. *J Food Sci. Technol* 32(4): 399-404.

Aucamp, M. C. ; Griff, J. T. ; Novellie, L. ; Papendick, B. ; Schwartz, H. M. and Steer, A. G. (1961). Kaffircorn malting and brewing studies. VIII. Nutritive value of some kaffircorn products. *J. Sci. Food Agric.* 12; 449-456.

Ayres, C. J. ; Mundt, O. J. and Sandline, E. W. (1980). *Microbiology of Foods.* W. H. Freeman, San Francisco, pp. 147-179.

Ayres, J. C. ; Mundt, J. O. and Sandine, W. E. (1980). *Microbiology of Foods* (edited by B. S. Schweigert). pp. 14-43, 147-159 , 180-205, 226-229. San Francisco,USA: W. H. Freeman.

Azhar, A. and Hamdy, M. K. (1981). *Biotechnol. Bioeng.* 23, 879.

Bachman, S. ; Gambus, H. and Nowotha, A. (1997). Effect of gamma-radiation on some physiochemical properties of triticale starch. *Pol. J. Food Nutr. Sci.* 6: 31-39.

Baervald, G. (1988). The origin bread based beverage. *Getreide Mhel Brot,* 42, 335-338.

Bao, J. and Corke, H. (2002). Pasting properties of ã-irradiated rice starches as affected by pH. *J. Agric. Food Chem.* 50: 336-341.

Bao, J. S. ; Shu, Q. Y. ; Xia, Y. W. ; Bergman, C. and McClung, A. (2001). Effect of gamma irradiation on aspects of milled rice (*Oryza sativa*) end-use quality. *J. Food Qual.* 24: 327-336.

Baranowski, J. D. and Nagel, C. W. (1981), Isolation and Identification of the Hydroxycinnamic Acid Derivatives in white Riesling Wine. *Am. J. Enol. Vitic.* 32, 5.

Basuki, T. (1977). The less well-known fermented foodstuffs of Indonesia. Symposium on indigenous fermented foods, Bangkok, Thailand.

Batra, L. R. (1981). Fermented cereals and grain legumes of India and vicinity. In M. Moo-Young. (ed.) Advances in Biotechnology, Vol. 11, Willowdale, Ontario, Canada: Pergamon press Canada, pp. 547-554.

Bayly, R. C. and Wigmore, G. J. (1973). Metabolism of phenol and cresois by mutants of *Pseudomonas putida. J. Bacteriol.* 113: 1112-1120.

Bergmeyer, H. (1974). *Methods in enzymatic analysis,* 2nd ed. Acadamic press, New York, N. Y.

Bernt E, Gutmann I. (1974). Ethanol determination with alcohol dehydrogenase and NAD. In: Bergmeyer HU (ed) *Methods of enzymatic analysis.* Academic Press, New York, pp 1499-1502.

Beuchat, L. R. (1995). Indigenous fermented foods. In: *Biotechnology : Enzymes, Biomass, Food and Feed* (edited by G. Reed and T. W. Nagodawithana). pp. 505-559. Weinheim, Germany : VCH.

Beyer, U. ; Tesche, M. ; Heller, W. and Sandermann, H. (1993). Fungistatic effectiveness of Phenolic compounds in Norway spruce (*Picea abies*) and the effect of SO_2. Forstwiss. Centralbl. 112, 251-256.

Bhattacharya, K. R. ; Sowbhagya, C. M. and Indudhara Swamy, Y. M. (1980). Quality of Indian rice. *J. Foods Sc. Techno.* 17: 189-193.

Bhella, R. S. and Altosaar, I. (1987). Translational control of α-amylase gene expression in *Aspergillus awamori. Biotechnol Appl.* Bio-Chem 9: 287-293.

Billek, G. and Kindl, H. (1962) Monastsh. *Chem.* 93, 85.

Bowles, B. L. and Miller, A. J. (1994). Caffeic acid activity against *Clostridium botulinum* spores. *J. Food Sci.* 59 :905 – 908.

Bray, H. C. and Thorpe, W. V. (1954). Analysis of Phenolic compound of interest in metabolism, *Meth Biochem*. Anal. 1: 27-32.

Brothwell, D. and Brothwell, P. (1969). *Foods in Antiquity*. Thomas and Hudson, London.

Cai, R. ; Heltiarachchy, N. S. and Jalaluddin, M. (2003). High-Performance Liquid Chromatography Determination of phenolic constituents in 17 varieties of Cowpeas. *J. Agric. Food Chem.* 51, 1623-1627.

Cain, R. B. (1980). The uptake and metabolism of lignin-related aromatic compounds and their regulation in microorganisms. P. 21-60. In T. K. Kirk, T. Higuchi, and H. M. Chang (ed), *lignin biodegradation: Microbiology, chemistry and potential application.* Vol. 1. CRC press Inc. , Boca Raton, Fla.

Campbell-Platt, G. (1994). Fermented Foods - a World Prospective. *Food Research International,* 27:253-257.

Cannel, E. and Moo-Young, M. (1980a). Solid state fermentation systems. *Pro Biochem.* 15:2-7.

Carlsen, M. (1994). *α-amylase production* by *Aspergillus oryzae.* Ph. D. thesis. Technical University of Denmark, **Lynby.**

Carlsen, M. and Nielson, J. (2001). Influence of carbon source on α-amylase production by *Aspergillus oryzae*. *Appl Microbiol Biotechnol.* 57: 346-349.

Carlsen, M.; Nielsen, J. and Villadsen, J. (1996a). Growth and á-amylase production by *Aspergillus oryzae* during continuous cultivations. *J Biotechnol* 45: 81-93.

Carlsen, M. ; Spohr, A. ; Nielsen, J. and Villadsen, J. (1996 b). Morphology and physiology of an á-amylase producing strain of *Aspergillus oryzae* during batch cultivations. *Biotechnol Bioeng* 49: 266-276.

Casey, G. P. and Ingledew, W. M. (1985). Ethanol tolerance in yeast. *Crit Rev Microbial* 13: 219-280.

Casey, G. P. ; Magnus, C. A. and Ingledew W. M. (1984). High gravity brewing: effects of nutrition on yeast consumption, Fermentative ability and alcohol production. *Appl. Environ Microbiol* 48:639-646.

Cha, Y. K. ; Kim, H. and Cadwallader, K. R. (1998). Aroma-active compounds in Kimchi during fermentation. *J. Agric Food Chem.* 46(5): 1944-53.

Chakraborty, T. K. ; Dwara Kanath, K. R. and Bhat, B. P. (1972). Studies on the Physio Chemical properties of the properties of some varieties of rice. *J. Foods Sci. Technol* 9 :140- 143.

Champ Szylet, M. ; Raiband, O. and Abdelkader, P. A. (1983). Amylase production by three lactobacillus strains isolated from chick peas. *Crop. J. Appl. Bacteriol.* 55: 487-493.

Cheig, H. S. and Park, K. Y. (1994). Biochemical microbiological and nutritional aspects Kimchi (Korean fermented vegetable products) *Crit rev Food Sci Nutr,* 34 (2): 175-203.

Chiao, J. S. (1981). Modernization of traditional Chinese fermented foods and beverages. In: *Advances in biotechnology*. Vol. 11, Edited by M. Moo-Young. Willowdale, Ontario, Canada: Pergamon press Canada, pp. 511-516.

Chopra. S. L. and Konwar, J. S. (1976). *Analytic Agricultural Chemistry*. Kalyani Pub. Ludhiana.

Chou, C. C. and Hwang, G. R. (1989). Fermentation of Tou-pan-chiang, a Chinese semi-solid condiment. Abstracts of papers read at a symposium on application of solid state fermentation of food and environmental Biotechnology Dec. 12-15, 1989, Taipei, Taiwan, Roc, P. 13.

Christensen, C. M. (1995). *The Moulds and Man*. 3rd Ed. Univ Minnessota Press, pp. 15-18.

Chung , J. H. (1967). Studies on the identification of organic acid and sugars in the fermented mash of the Takju made from different raw materials *J. Korean Agric. Chem. Soc.* 8 : 39-43.

Cook, P. E. and Campbell – Platt, G. (1994). *Aspergillus* and fermented foods. In: Powell, K. A. ; Renwick, A and Peberdy, J. F. editors. The genus *Aspergillus*. New York: Plenum Press. pp. 171-188.

Cronk, T. C.; Strienkraus, K. H. ; Hackler, L. R. and Mattick, L. R. (1977). Indonesian Tape Ketan fermentation. *Appl. Environ. Microbiol.* 33; 1067-1073.

Cronk,T. C. ; Mattick, L. R. Steinkraus, K. H. and Hackler, L. R. (1979). Production of higher alcohols during the Indonesian tape Ketan fermentation. *Appl. Environ. Microbial.* 37: 892-896.

Curic, D.; Dugum, J. and Bauman, I. (2002). The influence of fungal α-amylase supplementation on amylolytic activity and baking quality of flour. *International Journal of Food Science and Technology*. 37: 673-680.

Cutler, H. C. and Cardenas, M. (1947). Chicha, a native South American beer. *Bot. Mus. Leafl.* Harv. Univ. 13; 33-60.

D'Amore, T. and Stewart, G. (1987). Ethanol tolerence of yeast. *Enzyme Microb. Technol.,* 9: 322 – 330.

Dahiya, D. S. and Prabhu, K. A. (1977). *Indian jackfruit wine.* Symposium on Indigenous fermented foods, Bangkok, Thailand.

Dartey, C. K and Kinsella, J. E. (1973). *J. Agr. Food Chem.* 21; 721

De Wulf, O. ; Thonart, P. ; Gaignage, P. ; Marlier, M. ; Paris, A. and Paquot, M. (1986). Bioconversion of vanillin to vanillyl alcohol by *Saccharomyces cerevisiae.* Biotechnol. *Bioeng Symp* 17: 605-616

Deka, S. C. ; Sood, D. R. and Gupta, K. R. (2000). Nutritional evaluation of Basmati rice (*Oryzae sativa* L.) Genotypes. *J. Food Sci. Technol.* 37(3): 273-276.

Desta, B. (1977). A survey of the alcohol content of traditional beverages. *Ethiopian Med. J.* 15: 65-68.

Dirks, U. and Herrmann, K. (1984). *Phytochemistry* 23:1811.

Dirks, U. and Herrmann, K. (1984). Z. Lebensm. *Unters. Forsch.* 179, 12.

Djien, K. S. and Hesseltine, C. W. (1961). Indonesian fermented foods soybean *Dig.* 22: 14-15.

Dubois, M.; Gilles, K. ; Hamilton, J. K. ; Rerers, P. A. and Smith, F. (1951). *A colorimetric method for the determination of sugars.*

Durkee, A. B. (1977). *J. Agric. Food Chem.* 25: 286.

Durkee, A. B. and Thivierge, P. A. (1975). Bound phenolic acid in Brassica and Sinapsis Oilseeds. *J. Food Sci.* 40-820

Dutta, L. and Baruah, J. N. (1978). Nutrient composition of glutinous and non-glutinous rice varieties grown in Assam, India. *J. Agric . Sci.* 48 (10): 610 - 613.

Ekundayo, J. A. and Carlile, M. J. (1964). The germination of sporangiospores of *Rhizopus arrhizus;* Spore swelling and germ tube emergence. *J. Gen. Microbiol.* 35 : 261.

Ekundayo, J. A. (1977). *Nigerian pito.* Symposium of Indigenous fermented Foods, Bangkok, Thailand.

Ellis, J. J. ; Rhodes, L. J. and Hesseltine, C. W. (1976). The genus Amylomyces. *Mycologia* 68; 131-142.

Ellnain-Wojtaszek, M. and Zgõrka, G. (1999). High performance liquid chromatography and thin layer chromatography of phenolic acids from Ginkgo biloba L. leaves collected within vegetative period. *J. Liq. Chrom. Relat. Technol.* 22, 1457-1471

Erratt, J. A. ; Douglas, P. E. ; Moranelli, F. and Seligy, V. L. (1984). The introduction of á-amylase by starch in *Aspergillus oryzae*: evidence for controlled mRNA expression. *Can J. Biochem cul Biol* 62: 678-690.

Escobar, A. (1977). The South American maize beverage chicha. Thesis completed as part of the requirements for a master of Food Science Degree. Cornnell University, Ithaca, N. Y.

Escobar, A. ; Gardner, A. and Steinkraus, K. H. (1977). Studies of South American chicha. Symposium of Indigenous fermented foods, Bangkok, Thailand.

Esteve–Zarzoso. B. ; Peris–Toran, M. J. ; Garcia–Maiquez, E. ; URUBURU, F. and Querol , A. (2001). Yeast population dynamics during the fermentation and Biological Aging of Sherry Wines. *Applied and Environmental Microbiology.* p. 2056-2061.

Esteve-Zarzoso, B. ; Gostincar, A. ; Bobet, R. ; Uruburu, F. and Querol, A. (2000). Selection and molecular characterization of wine yeasts isolated from the "El penedes" alea (spain). *Food Microbiol.* 17: 553-562.

Ethiraj, S. and Suresh, E. R. (1992). Production of alcohol and vinegar from fermented Mango. *J. Food Science and Technol.* 29: 48-50.

Faparusi, S. I. (1970). Sugar changes during preparation of burukuku beer. *J. Sci. Food Agric.* 21; 79-81.

Fernandez–Flores, E. ; Kline, D. A. and Johnnson, A. R. (1970). *JAOAC* 53, 17-20.

Fernandez–Flores, E. ; Kline, D. A. ; Johnson, A. R. and Leber, B. L. (1970). *JAOAC* 53: 1200-1208.

Fitzpatrick, D. F. ; Bing, B. and Rohdewald, P. (1998). Endothelium-dependent vascular effects of pycnogenol. *J. Cardiovasc.* Pharmacol. 32, 509-515.

Fleet, G. H. ; Lafon-Lafon-Lafourcade, S. and Ribereau-Gayon, P. (1984). Evolution of yeast and Lactic acid bacteria during fermentation and storage of Bordeaux wines. *Appl. Environ. Microbiol.* 48:1034-1038.

Floridi, S.; Motanari, L. ; Marconi, O. and Fantozzi, P. (2003). Determination of free phenolic acids In wort and beer by colorimetric array detection. *J. Agric. Food Chem.* 51,1548-1554

Frazier, W. C. (1967). *Food Microbiology,* 2nd ed. McGraw-Hill Book Company, New York.

Frazier, W. C. (1967). "Food Microbiology", p. 413, McGraw -Hill, New York.

Garcia Maiquez, E. (1995). Los Microorganisms del Jerez. *Microbiol. SEM* 11: 51-58.

Gervais, P. and Sarrette, M. (1990). Influence of age of Mycelium and water activity of the medium on aroma production by *Trichoderma viride* grown on solid substrate. *Ferment Bioeng* 69 (1) : 46-50.

Glildyal, N. P. ; Lonesane, B. K. ; Sreekantiah, K. R. and Sreenivasa Muthy, V. (1985). Economics of submerged and solid state fermentations as for the production of amyloglocosidase. *J Food Sci Technol* 22: 171-176.

Gomes Da Silva, M. and Chaves Das Neves, H. (1997). Differentiation of strawberry varieties through purge-and-trap HRGC-MS, HRGC-FTIR and principal component analysis. *J. High. Rosolut. Chromatogr.* 20: 275-283.

Gomes Da Silva, M. and Chaves Das Neves, H. (1999). Complementary use of hyphenated purge-and trap gas chromatography techniques and sensory analysis in the aroma profiling of strawberries (*Fragaria ananassa*). *J. Agric. Food Chem.* 47: 4568-4573.

Gotcheva, V. ; Pandiella, S. S. ; Angelov, A. ; Rashkova, Z. and Webb, C. (2001). Monitoring the fermentation of the traditional Bulgarian beverages boza, *International Journal of Food Science and Technology* 36: 129-134.

Gromova, A. S. ; Lutskii, V. I. and Tyukavkina, N. A. (1977). *Chem. Nat. Comp.* (USSR) 13: 237.

Grover, R. K. (1964). The effect of amino acids on growth and sporulation of *Aspergillus flavus* and their carry-over for subsequent spore germination. *New Phytol.* 63:12.

Gsracheva, I. M. ; Zhirova, V. V. ; Babaeva, S. A. ; Kovalevich, L. S. and Makeev, D. M. (1972). Effect of carbohydrate composition of the fermentation medium on the synthesis of volatile acids by the yeast *Saccharomyces carlbergensis* 776, *Prikl. Biokhim, Mikrobiol.* 14: 583-585.

Hancioglu, O. and Karapinar, M. (1996). Microflora of boza, a traditional fermented Turkish beverages. *International Journal of Food Microbiology,* 35: 271-174.

Hang Y. D. and Lee Cywoodams. (1992). Production of alcohol from apple pomagi and damage guava or banana fruits. *J. Food Science and Technol.* 29: 48-50.

Harbone, J. B. (1989). Plant Phenolics. In methods of Plant Biochemistry; Dey, P. M. , Harbone, J. B. , Eds.; Academic Press: London, U. K. pp 1-27, 197-232.

Harborne, J. B. and Corner, J. J. (1961). *Biochem. J.* 81: 242.

Harborne, J. B. andWilliams, C. A. (1971). *Phytochemistry* 10: 2727.

Harkishore, K. M. (1977). *Kenyan sugarcane wine-muratina.* Symposiun on Indeginous Fermented Foods, Bangkok, Thailand.

Hasseltine, C. W. (1977a). Solid state fermentation. Part 1. *Proc. Biochem.* 12: 24-27.

Hasseltine, C. W. (1977b). Solid-state fermentation Part 2. *Proc Biochem* 12:29-32.

Hayashida, S. and Ohta, K. J. (1981). *Inst. Brew.* 42-49.

Hayashida, S. ; Kamachi, T. and Hongo, M. (1972). Semi continuous sake fermentation. *In Fermentation Technology Today,* edited by G. Terui. Osaka: Society of Fermentation Technology, Japan, pp. 645-649.

Heard, G. M. , and Fleet, G. H. (1985). Growth of natural yeast flora during the fermentation of inoculated wines. *Appl. Environ. Microbiol.* 50: 727-728.

Heermann, K. (1978). Progr. Chem. Org. Nat. Prod. 35: 73.

Hermann, K. (1989). Occurance and content of hydroxycinnamic and hydroxybenzoic acid compounds in foods. *Crit. Rev. Food Sci. Nutr.* 28 (4), 315-357

Hesseltine, C. W. ; Smith, M. ; Bradle, B. and Djien, K. S. (1963). Investigation of Tempeh, an Indonesian food, *Develop. Ind. Microbiol.* 4: 275.

Hesseltine, C. W. (1979). Some important fermented foods of mid-Asia, The middle East, and Africa. *J. Am. Oil Chem. Soc.* 56: 367-374.

Hesseltine, C. W. and Wang, H. L. (1967). *Traditional Fermentation Foods.* 9: 275-288.

Hesseltine, C. W. and Wang, H. L. (1967). *Biotechnol. Bioeng.* 9, 275.

Hesseltine, C. W. (1965). A millennium of fungi, food and fermentation. *Mycologia* 57(2):149-197.

Hesseltine, C. W. ; Featherton, C. L. ; Lomband, G. L. and Dowell, V. R. (1985). Anaerobic growth of moulds isolated from starters for foods in Asian countries. *Mycologia* 779(3): 390-400.

Hesseltine, C. W. ;Smith, M. and Wang, H. L. (1967). New fermented cereal products, *Develop. Ind. Microbiol.* 8 : 178.

Hirayama, K. K. ; Tobita, S. and Hirayama, K. (1994). Biodegradation of phenol and monochlorophenols by Yeaast *Rhodotorula glutinis. Water Sci. Technol;* 30: 59-66.

Hong, S. W. ; Hah, Y. S. and Lim, B. J. (1968a). The chemical components and their changes during and after brewing maggally. *Report of the Brewing Lab. Of the National Tax office I.* Seoul. Korea.

Hoover, R. (2000). Acid –treated starches. *Food Rev. Int.* 16, 369-392.

Hoover, R. and Sailaja, Y. (1996). F. W. Characterization of starch from wild and long grain brown rice. *Food Res. Int.,* 29:99-107.

Huang, W. and Chan, J. (1982). Studies on decomposition of phenol by immobilized cells of *Candida tropicalis*. No. 314 *Huangjeng Kexue Zuebae,* 2: 293-298.

Hughes, E. J. L. ; Bayly, R. C. and Skurray, R. A. (1984). Evidence for isofunctional enzyme in the degradation of phenol, m- and p- toluate, and p-cresol via Catechol meta-cleavage pathways in Alcaligenes eutrophus. *J Bacteriol* 158: 79-83.

Hutchingson, C. M. and C. S. Ram-Ayyar. (1925). Bakhar - The Indian rice ferment. *Mem. Indian Dept. Agric. Bact.* 1:137-168.

Hwang, G. R. (1988). Studies on the fermentation of Tou-Pan-Chiang- A Chinese Traditional fermented food. Ph.D. Thesis, Graduate Institute of Food Science and Technology, National Taiwan University Taipei, Taiwan ROC.

Ibeas, J. I. ; Lozano, I. ; Perdigones, F. and Jimenes. J. (1997). Dynamics of flor yeast populations during the biological aging of sherry wines. *Am. J. Enol. Vitic.* 48:75-79.

Iland, P. ; Ewart, A. and Sitters. (1993). J. Techniques for chemical analysis and stability test of Grape Juice and Wine; Patric Iland Wine production: Campbell Town South Australia.

Ingram, L. O. (1986). *Trends Biotechnol.* 4: 40 – 44.

Ingram, L. O. and Buttke , T. M. (1984). *Adv. Microb. Physiol.* 25 : 253-300.

Ingram, L. O. ; Dombek, K. M. and Osman, Y. A. (1986). *SIM News* 36 : 8 –11.

Ito, A. ; Kasai, R. ; Yamasaki, K. and Sugimoto, H. (1993). *Phytochemistry.* 33; 1133.

Ito, A. ; Kasai, R. ; Yamasaki, K. ; Duc, N. M. and Nham, N. T. (1994). LIGNAN GLYCOSIDES FROM BARK OF

ALBIZZIA MYRIOPHYLLA. *Phytochemistry.* 37(5), pp. 1455-1458.

Ivanova, S. Z.; Medvedeva, S. A. ; Tyukavkina, N. A. and Zelenikina, N. D. (1975). *Chem. Nat. Comp.* (USSR) 11,424.

Izuka, H. and Lin, C. F. (1981). On the genus monascus of Asia and its specific characteristics. *In Advances in Biotechnology.* Vol. 11, edited by M. Moo-young, Willowdale, Ontario and Canada: Pergamon press Canada, pp. 555-562.

Jones, K. H.; Trudgill, P. W. and Hopper D. J. (1993). Metabolism of p-Cresol by the fungus *Aspergillus fumigatus. Appl. Environ. Microbiol.* 59: 1125-1130.

Jones, K. H.; Trudgill, P. W. and Hopper, D. J. (1995). Evidence of two pathways for the metabolism of phenol by *Aspergillus fumigatus. Arch: Microbiol;* 27:441-447.

Jones, R. P.; Pamnent, N. and Greenfield, P. F. (1981). Alcohol fermentation by yeast- the effect of environmental and other variables. *Process Bio chem.* 16:42-49.

Juliano, B . O.; Onate, L. U. and Lu Mundo, A. M. (1965). Relation of starch composition and protein content and gelatilization temperature to cooking and eating qualities of milled rice. *Food Technol.* 19 : 106.

Juliano, B. O. ; Bantista, G. M. ; Dugay, J. C. and Rayes, A. C. (1964). Studies on the physiochemical properties of rice. *J. Agric. Food Chem.* 12 : 121-138.

Juliano, B. O. (1972). In: Rice Chemistry and Technology. Houston, D. F. (ed.) AACC St. *Paul. Minn.* Chapter - 2, pp. 16-74.

Juliano, B. O. (1985). *Rice Chemistry and Technology* AACC. St Paul MN.

Juliano, B. O. (1972). The rice caryopsis and its composition. In: Houston, D. F. (ed). Rice: Chemistry and Technology. *Am. Assoc. Cereal chem.,* st. Paul, MN.

Juliano, B. O. and Boulter, D. (1976). Extraction and composition of rice endosperm glutelin. *Phytochemistry.* 15: 1601.

Kahkonan, M. P. ; Hopia, A. I. ; Vuorela, H. J; Rauha, J. P. ; Pihlajah, K. ; Kujala, T. S. ; Heinonen, M. and Heinonen, M. (1999). Antioxidant activity of plant extracts containing phenolic compounds. *J. Agric. Food Chem.* 47, 3954-3962.

Kang, I. J. ; Chung, C. K. and Sohn, J. I. (1999). Effects of pH and gamma irradiation on the physiochemical properties of corn starch. *J. Food Sci. Nutr.* 175-179.

Kang, J. H. ; Lee, J. H. ; Min, S. and Min, D. B. (2003). Changes of volatile compounds, Lactic Acids Bacteria, pH and Headspace Gases in Kimchi a traditional Korean Fermented vegetable product. *Journal of Food Science.* 68, Nr. 3: 849-854.

Khetarpaul, N. and Chauhan, B. M. (1991) . Sequential fermentation of pearl millet by yeasts and lactobacilli-changes in available carbohydrates content. *Food Chemistry,* 40 : 235 – 240.

Kim, C. J. (1968). Microbiological and enzymological studies on Takju brewing. *J. Korean Agric. Chem. Soc.* 10: 69-99.

Kim, C. J. (1970). Studies on the differentiation and the population changes of Takju Yeasts by the TTC-agar overlay techniques. *Korean. J. Microbiol.* 8: 69-76.

Kim, J. O. and Lee, B. H. (1970). Taxonomical studies of yeasts in Korea. I. on yeasts isolated from Takju. *Korean J. Microbiol.* 8: 77-84.

Kinsella, E. and Hwang, J. D. (1976). Biosynthesis of flavors by *Penicillium queforti. Biotechnology and Bioengineering.* 18: 927-938.

Ko, S. D. (1972). Tape fermentation. *Appl. Microbiol.* 23:976-978.

Ko, S. D. (1977). Indonesian Ragi. Symposium on indegenous fermented foods, Bangkok, Thailand.

Kodama, K. (1986). Microbial interaction in Japanese sake fermentation. In *Perspective in Microbial Ecology*, edited by F. Mecuiar and M. Gantar. Ljubljana, Yugoslavia: Solvene Society of Microbiology, pp. 289-298.

Kodama, K. and Yoshizawa,K. (1977). Sake. In *Economic Microbiology* (A. H. Rose, ed.), vol. I. Academic Press, N.Y., pp. 432-475.

Koksel, H. ; Celik, S. and Tuncer, T. (1996). Effects of gamma irradiation on durum wheat and spaghetti quality. *Cereal chem.* 73, 506-509.

Kolhekar, S. R. ; Mahajan, P. B. ; Ambedkar, S. S. and Borker, P. S. (1985). Purification and characterization of glucoamylase from a higher yielding mutant of *Aspergillus candidus* link var. aureus. *Appl Microbiol Biotechnol* 22: 181-186.

Konishi, Y. ; Wakai, Y. ; Maenaka, M. and Shiraisi, M. (1989). Cold-growing *Pseudomonas spp.* in the early stage of yeast fermentation starter mash (Yamahaimoio). A traditional sake brewing. *Abstract* of papers read at 5th international symposium on microbial Ecology. Aug. , Kyoto, Japan, p. 97.

Kozaki, M. (1976) . Fermented foods and related Microorganisms in Southeast Asia. *Proc. Jap. Assoc. of Mycotoxicology* 2 : 1 – 9 .

Kreger-van Rij, N. J. W. (1984). *The yeasts; a taxonomic study,* 3rd ed. Elsevier, Amsterdam, The Netherlands.

Kryggier, K.; Sosulski, F. and Hogge, L. (1982). Free, Esterified and insoluble phenolic acids. I. Extraction and purification procedure. *J. Agric. Food Chem.* 30,330-334.

Kushner, D. J. and Pollock, M. R. (1961). The location of cell- bound penicillinase in *Bacillus subtilis. J. Gen. Microbiol.* 26: 255-265.

Lachmund, A. ; Urmann, U. ; Minol, K. ; Wirsel, S. and Ruttkowski, E. (1993). Regulation of α-amylase formation in *Aspergillus oryzae* and *Aspergillus nidulans* transformants. *Curr. Microbiol* 26: 47-51.

Lai, M. N. (1984). Studies on solid state fermentation of Sorghum Liquor with heat transfer model. Ph.D. Thesis, Graduate Institute of Agricultural chemistry, National Taiwan University, Taipei, Taiwan, ROC.

Lai, M. N. ; Wang, H. H. and Chang, F. N. (1989). Thermal diffusivity of solid mash of sorghum brewing- A solid state fermentation. *Biotechnol. Bio engrg.* 34:1337-1340.

Lea, A. G. H. and Arnold, G. M. (1978). The phenolic of ciders. bitterness and astringency. *Journal of the Science of Food and Agriculture,* 29: 178-463.

Lee , Z. S. and Rhee, T. W. (1970). Studies on the Microflora of Takju brewing. *Korean J . Microbiol.* 8 : 116 – 133 .

Lee, J. H. ; Kang, J. H. and Min, D. B. (2003). Optimization of solid-phase micro extraction for the Analysisof the Headspace volatile compounds in Kimchi, a traditional Korean Fermented vegetable product. *Journal of Food Science.* 68, Nr 3: 844-848.

Lee, S. B. ; Chang,W. G. ; Lim,B. J. and Kim,D. C. (1969). Studies on chemical components of fermented mash in the brewing of maggally. *Korean J. Microbiol.* 7;153-158.

Lee,S. Y. ; Shin, Y. C. ; Lee, S. H. ; Park, S. S. ;Kim, H. S. and Byun, S. M. (1984). *Korean J. Food Sci. Technol,* 16(4), 463.

Lin, C. F. ; Chang, C. C. and Lu, S. S. (1975). Taxonomy study on the *Rhizopus sp.* isolated from starter of shao-hsing wine. *J. Chin. Agric. Chem. Soc.* 13(1-2): 81-87.

Lin, Y. –S. ; Tsai, Y. –J. ; Tsay, J. –S. and Lin, J. -K. (2003). Factors affecting the level of tea polyphenols and Caffeine in tea lesves. *J. Agric. Food Chem.* 51, 1864-1873

Liu, P. W. ; Chen, S. H. and Chen, C. C. (1959c). Studies on Shao-hsing wine 111. on the pilot-plant experiment of Lin-fan wine. *Chemistry* 48(4): 241-244 (in Chinese).

Liu, P. W. ; Chen, S. H. ; Chiu, T. H. and Chen, C. C. (1959a). Studies on shao-hsing wine. *Chemistry* 48(3): 118-188 (in Chinese).

Liu, P. W. ; Chen, S. H. ; Chiu, T. H. ; Young, C. and Huang, Y. Y. (1959b). Studies on Shao-hsing wine 11. on the manufacture of Tein-chiu-niang and Lin-fan wine. *Chemistry* 48 (3): 189-195 (in Chinese).

Liu, P. W. ; Chen, S. H. ; Lee, T. H. and Chang, K. T. (1966). Studies on Shao-hsing wine IV. On the pilot-plant experiment of Lin-fan wine. *Chemistry* 49(3): 155-159 (in Chinese)

Lizuka, M. ; Uenakai, K. ; Svendsby, O. and Yamamoto, T. (1985). Alcohol Fermentation of Green Banana. *J. Ferment.* Technol; 63(5) 475-477.

Lovelace, C. E. A. (1977). Estimation of nutrient content of two fermented beverages from Zambia, opaque maize beer and munkoyo. Symposium on Indigenous Fermented Foods, Bangkok, Thailand.

Lowery, O. H. ; Rosenbrough, N. J. ; Farr, A. L. and Randall, R. J. (1951). Protein measurement with the folair phenol reagent. *J. Biol. Chem.* 193: 256-275.

Lundgren, L. (1987). Studies on low-molecular Phenolic constituents in conifer needles. Dissertation, Swedish University of Agricultural Science, Uppsala, Sweden.

Lynch, B. T. ; Glass, R. L. and Gedder, W. F. (1962). Grain storage studies of quantitative changes occurring in the sugar of wheat deterioration in the presence and absence of moulds. *Cereal Chem.* 39: 256-262.

Mac Arthur, L. A. and D'Appolonia, B. L. (1984). Gamma radiation of wheat. II. Effects of low-dosage radiations on starch properties. *Cereal Chem.* 61: 321-326.

Maillord, M. N. and Besset, C. (1995). Evolution of antioxidant activity during kilning: role of insoluble bound phenolic acids of barley and malt. *J. Agric. Food Chem.* 43, 1789-1793

Maiorella, B. ; Blanch, H. W. and Wilke, C. R. (1983). By product inhibition effect on ethanolic fermentation by *Saccharomyces cerevisiae. Biotechnol. Bioeng.* 25: 103-125

Malamy, M. and Horecker, L. (1961). The localization of alkaline phosphatase in E. coli K_{12} Biochem. *Biophys. Res. Commun.* 5: 104-108.

Marrinez, P. ; Condon, A. C. ; Perez, L. and Beniez, T. (1995). Physiological and molecular characterization of flor yeasts: polymorphism of flor yeast populations yeast 11: 1399-1411.

Martinelli, A. F. and Hesseltine, C. W. (1964). Tempeh fermentation : Package tray fermentations. *Food Technol.* 18 :167.

Martinez de la ossa, E. ; Perez, L. and Caro, I. (1987). Variations of the major volatiles through aging of Sherry *Am. J. Enol. Vitic.* 38 : 293 – 297.

Martinez, P. ; Perez Rodriguez, L. and Benitez, T. (1997). Evolution of flor yeast population during the biological aging of fine sherry wine. Am. *J. Enol. Vatic.* 48: 160-168.

Maxwell, T. J. (1956). Agricultural ceremonies of the central Andes during four hundred years of Spanish contact. *Ethnohistory* 3: 46-71.

McCall, E. R. ; Turgens, J. F. ; Hoffpauir, C. L. ; Pon, W. A. ; Jr. , Stark, S. M. ; Jr; CUCULLU, A. F. ; Heinzlman, D. C. ; Cirino, V. O. and Murray, M. D. (1953). Composition of rice: Influence of variety and environment on physical and chemical composition. *J. Agric. Food Chem.* 1: 988.

Merican, Z. (1977). *Malaysian coconut palm today.* Symposium on Indigenous Fermented Foods, Bangkok, Thailand.

Michel, J. P. ; Raffi, J. ; Saint, Lebe, L. ; Huchette, M. and Fleche, G. (1980). Experimental study of the depolymerization of starch under the combined action of protons and gamma radiation. *Starch/staerke* 32: 340-344.

Middlehoven, W. J. ; Coenen, A. ; Kraakman, B, and Geipke, M. D. S. (1992). Degradation of some phenols and hydroxybenzoates by the imperfect Ascomycetous yeast *Candida parapsilosis* and *Arxula adeninivorans*: evidence for an operative gentisate pathway. *Antonio Van Leeuwenhock* 62: 181-187.

Mizobuchi, S. and Sato, Y. (1984). *Agric. Biol. Chem.* 48, 2771-2775.

Morcas, S. R. (1977). *Egyptian bouza.* Symposium on Indigenous Fermented foods, Bangkok, Thailand.

MØrkeberg, R. ; Carlsel, M. and Nielsen, J. (1995). Introduction and repression of α-amylase production in recombinant and wild-type strains of *Aspergillus oryzae. Micrbiology* 141: 2449-2454.

Morse, R. A. and Steinkraus, K. H. (1975). Wines from the fermentation of the honey. Chapter 16. In *Honey, a comprehensive survey* (Eva Crane, ed.). Heinemann, London, pp. 392-407.

Murakami, H. (1972). Some problem in sake brewing. In *Fermentation Technology Today*, edited by G. Terui. Osaka: Society of Fermentation Technology, Japan, pp. 639-643.

Nago, M. C. ; Hounhouigan, J. D. ; Zanou, E. and Mestres, C. (1998). Characterization of the Beninese traditional ogi, a fermented maize slurry :Physicochemical and Microbiological aspects. *International Journal of Food Science and Technology.* 33, 307-315.

Nakano, M. ; Ebine, H. and Ito, H. (1976). On the miso: A Japanese Traditional fermented food, In view of ecological system of fermentation microorganisms depending on the types of miso products. *Abstract* of papers read at 5th International fermentation symposium, June 28- July 3, 1976, at Berlin, p. 363.

Narayan, R. ; Chauhan, G. S. and Verma, N. S. (1988). Changes in the quality of soyabean during storage. Part 1. Effect of storage of some physico chemical properties of Soyabean. *Food Chem.* 27:13-23.

Nelson, N. (1944). A photometric adaptation of the Somogyi method for the determination of glucose. *J. Biol. Chem.* 153: 375-380.

Neu, H. C. and Heppel, L. A. (1964). On the surface localization of enzymes in E. coli. *Biochem. Biophys. Res. Commun.* 17: 215-219.

Neujahr, H. Y. and Varga, J. M. (1970). Degradation of phenols by intact cells and cell-free preparations of *Trichosporon cutaneum. Eur. J. Biochem* 13: 37-44. New york, N. Y.

Nice, D. J. ; Robinson, D. S. and Holden, M. A. (1995). Characterization of heat-stable antioxidant co-purified with the super oxide dismutase activity from dried peas. *Food Chem.* 52 (4), 393-397.

Nicholson, G. E. (1960). Chicha maize types and Chicha manufacture in Peru. *Econ. Bot.* 14: 209-299.

Niemann, G. J. (1969). *Phytochemistry.* 8,2101.

Niemann, G. J. (1973). *Phytochemistry.* 12,723.

Niemann, G; J. and Bass, W. J. (1978). *Z. Naturforsch.* 33C; 780.

Noda, Y. ; Anzai, K. ; Mori A. ; Kohno, M. ; Shinmei, M. and Packer, L. (1997). Hydroxyl and superoxide anion radical scavenging activities of natural sources antioxidants using the computerized JES-FR30 ESR spectometer system. *Biochem. Mol. Biol. Int.* 42, 35-44.

Novelie, L. and Schutte, R. J. (1961). Kaffircorn malting and brewing studies. X. The susceptibility of sorgham starch to amylolysis. *J. Sci. Food Agric.* 12: 552-59.

Novellie, L. (1960a). Kaffircorn malting and brewing studies. IV. The attraction and nature of the unsoluble amylases of Kaffircorn malts. *J. Sci. Food Agric.* 11: 408-421.

Novellie, L. (1960b). Kaffircorn malting and brewing studies. V. Occurrence of â-amylase in Kaffircorn malts. *J. Sci. Food Agric.* 11: 457-471.

Novellie, L. (1966b). Kaffircorn malting and brewing studies. XIV. Mashing with Kaffircorn malt. Factors affecting sugar production. *J. Sci. Food Agric.* 17: 552-559.

Nunokawa, Y. (1981). Enzymes affecting the fermentation of sake moromi-mash. In *Advances in Biotechnology,* edited by M. Moo-young. Canada Pergamon press Canada, pp. 562-568.

Nyako, K. O. (1977). *Palm-Wine-an alcohol beverage of Ghana.* Symposium on Indigenous Fermented Foods, Bangkok, Thailand.

Nykanen, I. (1986). Formation and occurrence of flavor compounds in wine and distilled alcohol beverages. *American Journal of Enology and Viticulture,* 37: 84-96.

O' Donovan, M. N. and Novellie, L. (1966). Kaffircorn malting and brewing Studies. XV. The fuel oils of Kaffir beer. *J. Sci. Food Agric.* 17: 362- 365

Ogawa, M. and Ogilara, Y. (1976). *Chem. Pharm. Bull.* 24, 2102.

Okafor, N. (1972b). The source of microorganisms in Palm-Wine. Symposium proceedings. Nigerian Society for Microbiology. 1: 102-106.

Okafor, N. (1977b). The microbiological basis of a method for palm-wine preservation. *J. Appl. Bacteriol.* 43: 159-161.

Okofor, N. (1972a). Palm-Wine Yeasts from parts of Nigeria. *J. Sci. Food Agric.* 23: 1399-1407.

Oriol, E. ; Schettino, B. ; Viniegra–Gonzales, G. and Raimbault, M. (1988). Solid state culture of *Aspergillus niger* on support. *J Ferment Technol* 66(1) : 57-62.

Osman, Y. and Ingram, L. O. J. (1985). *Bacteriol.* 164, 173 – 180.

Ossipov, V. ; Nurmi, K. ; Loponen, J.; Haukioja, E. and Pihlaja, K. (1996). HPLC seperation and identification of Phenolic compounds from leaves of Betula pubescens and Betula pendula. *J. Chromatogr.* A 721, 59-68.

Ottogalli, G. and Galli, A. (1997). Fermented foods in the past and future. Annali Di Microbiologia ed Enzimologia, 47, 237-257.

Pan, H. (1995). Studies on phenolic and terpenoid constituents extracted from bark of birch, Spruce and Pine in Sweden. Dissertation Swedish University of Agricultural Sciences, Uppsala, Sweden.

Panchal C. J. and Stewart, G. G. (1981). *Dev. Ind. Microbial.* 22: 711-717.

Panchal, C. J. and Stewart, G. G. (1981). Ethanol production by a highly flocculent brewing yeast strain *Dev. Ind. Microbial* 22: 711-717.

Pandey, A. ; Ashkumary, L. ; Selvakumar, P. and Vjayalakshmi, K. S. (1994a). Influence of water activity on growth and activity of *Aspergillus niger* for glucoamylase production on solid state fermentation. *WJ Microbiol Technol* 10 :485-486.

Park, Y. S. ; Kang, S. W. ; Lee, J. S. ; Hong, S. l. and Kim, S. W. (2002). Xylanase production in solid state fermentation by *Aspergillus niger* mutant using statistical experimental designs. *Appl Microbial Biotechnol.* 58: 761-766.

Peach, K. and Tracey, M. V. (1956). *Modern method of plant analysis*, Vol. I: Springer Verlog, Berlin.

Peach, K. and Tracey, M. V. (1955a). *Modern method of plain analysis*, vol. II: Speinger Verlog, Berlin (1955a).

Peach, K. and Tracey, M. V. (1955b). *Modern method of plant analysis,* Vol. IV: Springer Verlog Berlin.

Pedersen, H. and Neilsen. (2000). The Influence of nitrogen sources on the α-amylase productivity of *Aspergillus oryzae* in continuous cultures. *Appl. Microbiol Biotechnol* 53: 278-281.

Pederson, C. S. (1979). *Microbiology of Food Fermentations*, 2nd ed . AvI Publishing Company, Westport, CT.

Pennington, C. W. (1969). *The Tepehuan of Chihuahua, Their Material culture.* Univ. of Utah Press, Salt Lake City, Utah, pp. 90, 107, 110.

Pichyangkura, S. and Kulprecha, S. (1977). Survey of Mycelial moulds in loogpang from various sources in Thailand. Symposium on indigenous fermented foods, Bangkok, Thailand.

Pietta, P. ; Mauri, P. and Rava, A. (1988). Reversed-phase high performance Liquid Chromatographic method for the analysis of biflavones in Ginkgo biloba L. extracts. *J. Chromatogr.* 437, 453-456.

Piga, A. ; Caro, A. and Corda, G. (2003). From Plums to Prunes: Influence of drying Parameters on Polyphenols and Antioxidant Activity. *J. Agric. Food Chem.* 51: 3675-3681.

Platt, B. S. (1955). Some traditional alcoholic beverages and their importance in indigenous African Communities . proc. *Nutr. Soc.* 14 : 115 – 124.

Platt, B. S. (1964). Biological ennoblement :improvement of nutritive value of foods and dictary regiments by biological agencies. *Food Technol.* 18 : 662 – 670 .

Pollock, J. R. A. (1979). *Brewing Science*. 1 Vol. , Academic press, London.

Pradhan. S. (1969). *Insect pests of crops*, National Book Trust, New Delhi, pp-208.

Prescott, S. C. and Dunn, C. B. (1959). *Industrial Microbiology*; 3rd Edn. McGraw Hill Inc. New York.

Pruthi, H. S. (1969). Pest of Agriculture commodities in storage. In: *Text book on Agricultural Entomology*, ICAR, New Delhi.

Pushpama, P. and Reddy Uma. M. (1979). Physico-chemical changes in rice and jowar stored in different agro-climatic regions of Andhra Pradesh. *Bull. Grain Tech.*, 27(2); 97-108.

Qi, B. ; Moc, W. M. and Kinney, K. A. (2002). Biodegradation of volatile organic compounds by five fungal species. *Appl. Microbial Biotechnol.* 58: 684-689.

Querol, A. ; Barrio, E.. and Ramon, D. (1994). Population dynamics of natural *Saccharomyces* Strains during wine fermentation. *Int. J. Food Microbiol.* 21 : 315 – 323.

Raffi, J. ; Michel, J. P. and Saint-Lebe, L. (1980). Theoritical study of the radio polymerization of starch. *Starch/staerke* 32: 227-229.

Raghavaiah, P. and Kaul, A. K. (1970). Chemical composition, cooking quality and protein content of some varieties of rice. *Indian J. Genet Plant Bred.* 30(3): 738-743.

Rayas-Duarte, P. and Rupnow, J. H. (1993). Gamma-irradiation affects some physical properties of dry beans (*Phascolus vulgaris*) Starch. *J. Food Sci.* 58: 389-394.

Rebbéreau-Gayon, P. (1972). *Plant phenolics* Oliver and Boyd Edinburgh p. 257.

Reinders, R. D. ; Biesterveld, S. and Bijker, P. G. H. (2001). Survival of *Escherichia coli* 0157 : H7ATCC 43 895 in a model apple juice medium with difference concentration of proline and caffeic acid. *Appl. Environ. Microbiol.* 67:2863 – 2866.

Ribereau-Gayon, P. (1972). *Plant phenolics,* chpt. 4. Hafner Publishing, New York.

Roushdi, M. ; Harras, A. ; El-meligi, A. and Bassim, M. (1983). Effect of high doses of gamma rays on corn grains. II Influence on some physical and chemical properties of starch and its fraction. *Strach/staerke.* 35:15-18.

Ryan, J. J. (1972). Chemical composition of Canadian Apple juice. *Journal of the AOAC* 55(5): 1105-1109.

Sabularse, V. C. ; Liuzzo, J. A. ; Rao, R. M. and Grodner, R. M. (1991). Cooking Quality of brown rice as influence by gamma-irradiation, variety and storage. *J. Food Sci.* 56; 96-98, 108.

Samjeewa, U. ; Pethiyagoda, U. ; Atputharajah, J. O. and Wijertna, M. C. P. (1976a). Fermentation characteristics of some yeast isolated from coconut Research and Develôpment, Kasaragod, India.

Sanchez, P. C. (1977). Phillippine sugarcane Wine-basi. Symposium on Indigenous Fermented Foods, Bangkok, Thailand.

Sanchez-Marroquin, A. (1953). The biochemical activity of some microorganisms of pulque. *Mem. Congr. Cient. Mex.* IV Centenario Univ. Mex. 2:471-484.

Sanchez-Marroquin, A. (1977). Mexican pulque-a fermented drink from agave juice. Symposium on Indigenous Fermented Foods, Bangkok, Thailand.

Sanchez-Marroquin, A. and Hope, P. H. (1953). Agave juice fermentation and chemical composition studies of some species. *Agric. Food Chem.* 1:246-249.

Sanni, A. I. (1993). The need for process optimization of African fermentedfoods and beverages. *International Journal of Food Microbiology,* 18 : 85 – 95.

Saona, S. ; Basuki, T. and Sastraatmadja, D. D. (1977). Indonesia ragi. Symposium on Indigenous Fermented foods, Bangkok, Thailand.

Saono, S. ; Gandjar, I. ; Basuki, T. and Karsono, H. (1974). Mycoflora of "ragi" and some other traditional fermented foods of Indonesia. *Annales Bororienses* V(4): 187-204.

Schuster, B. and Herrmann, K. (1985). *Phytochemistry*. 24: 2761.

Schuster, B. ; Winter, M. and Herrmann, K. (1986). *Z. Naturforsch.* 41C:511.

Schutz, M. and Gafner, J. (1993). Analysis of yeast diversity during spontaneous and induced alcoholic fermentations, *J. Appl. Bacteriol.* 75:511-558.

Senter, S. D. ; Horvat, R. J. and Forbus, W. R. (1983). *J. Food Sci.* 48: 798.

Seo, A. and Morr, C. V. (1984). Improved High-performance Liquid chromatographic Analysis of phenolic Acids and Isoflavonoids from soyabean protein products. *J. Agric. Food Chem.;* 32(3): 530-533.

Shahidi, F. and Naczk, M. (1992). An interview of phenolic of canola and rapeseed: chemical, sensory and nutritional significance. *J. Am. Oil Chem.* Soc. 69. 917-924

Shahrzad, S. and Bitsch, I. (1996). Determination of some pharmacologically active phenolic acids in juice by high-performance liquid chromatography. *Journal of Chromatography* A, 741: 223-231.

Shanmuganathan, V,; Arasaratnam, V. and Balasubramaniam, K. (1994). Optimization of culture conditions for the production of protease from *Aspergillus niger* by solid state fermentation. In Proceedings of 3rd Annual Sessions Jaffna science Association Srilanka P25

Sharma, H. R. and Bains, G. S. (1979). Studies on the variability in the quality of market paddy. *J. Food Sci. Technol.* 16: 67-69.

Shibasaki, S. K. and Hesseltine, C. W. (1962). Miso fermentation. *Econ. Bot.* 16 (3): 180-195.

Shima, K. ; Hisada, S. and Ingaki, L. (1971). *Phytochemistry.* 10:894.

Sie, S. N. (1962). *Ragi Baras,* unpublished M. S. Thesis, Bandung Institute of Technology, Indonesia.

Singh, G. (1987). Microbial and Chemical contain of the fermented bamboo shoot, "Soibum." *Frontr. Bot.* 1: 89-95.

Singh, N. ; Singh, H. ; Kaur, K. and Bakshi, M. S. (2000). Determining the effect of degree of polish and ash distribution in brown rice using conductivity. *Food Chem.* 69: 147-151.

Singh, R. (1989). Starch metabolisms in higher plants, In: *Recent Advances in plant Biochemistry.* Mehta, S. L. ; Lodha, M. L. and Sane, P. V. (ed.), ICAR, New Delhi, pp. 107-136.

Singleton, V. L. ; Timberlake, C. F. and Lea, A. G. H. (1978). The Phenolic cinnamates of white grape. *J. Sci. Food Agric.* 29, 403.

Slinkard, K. and Singleton,V. L(1977). Total phenolic analysis: Automation and comparision with manual methods. *Am. J. Enol. Vitic.* 28,49-55

Soedarsono, J. (1970). Some notes on "ragi tape," an Inoculum for "tape" fermentation. *II mu pertaniu* 1:235-241.

Sokhey, A. S. and Chinnaswasmy, R. (1993). Chemical and molecular properties of irradiated starch extrudates. *Cereal Chem.* 70: 260-268.

Sokhey, A. S. ; Hanna, M. A. ; Wootton, M. and Gallant, D. J. (1993). Properties of irradiated starches. *Food Struct.* 12: 397-410.

Somogyi, M. (1952). Notes on sugar determination. *J. Biol. Chem.* 195: 19-23.

Sone, K. (1959). The quality of plant food. 11. Influence of variety and environment upon the riboflavin content of hulled rice. 2. In special reference to its effective factors. *Tocklai J. Agric.* Res. 9: 179-197.

Stahl, E. (1969). *Thin-layer chromatography*, p. 857. Springer-Verlag, NewYork Inc. ,N. Y.

Stahl, G. (1966). *Thin layer chromatography*, Academy Press, New York.

Steinkraus, K. H. (1977). *Handbook of indigenous fermented foods* (edited by K. H. Steinkraus). pp. 6-12, 132 –134, 148 – 149, 179 – 195, 357 – 367. USA ;Marcel Dekker.

Steinkraus, K. H(1997) Classification of fermented foods: World Wide review of household fermentation techniques. *Food Control.* 8 : 311- 317.

Steinkraus, K. H. (1983) Traditional foods fermentation as Industrial resources, *Acta Biotecnologica.* 3.

Steinkraus, K. H. and Morse, R. A. (1966). Factors influencing the fermentation of honey in mead production. *J. Api. Res.* 5:17-26.

Stevens, J. F. ; Ivancic, M. ; Hsu, V. L. and Deinzer, M. L. (1997). Prenylflavanoids from *Humulus lupulus. Phytochemistry* 44, 1575-1585.

Stract, D. ; Heilemann, J. ; Wray, V. and Disks, H. (1989). Structures and accumulation patterns of soluble and insoluble phenolic from Norway Spruce needles. *Photochemistry.* 28, 2071-2078.

Straubs, G. (1987). Phenol hydroxylase from *Rhodococcus sp. pl.* J basic *Microbiol* 27: 229-232.

Swambhagya, C. M. and Bhattacharya, K. R. (1979). Simplified Determination of Amylose in Milled rice, *Strake*, pp. 159-163.

Swings, J. and Deley, J. (1977). The biology of *Zymomonas*. *Bact. Rev.* 41:1-46.

Taboada, J.; Ulloa, M. and Herrera, T. (1977). Microbiological studies on tesguino, a fermented maize beverage consumed in Northern and Central Mexico. Symposium on Indigenous Fermented Foods, Bangkok, Thailand.

Takagi, T. and Lida, T. (1980). Antioxidant for fats and oils from canary seed: sterol and triterpenealcohol esters of caffeic acids. *J. Am. Oil Chem. Soc.* 57, 326-329.

Tanimura, W.; Sanchez, P. C. and Kozaki, M. (1978b). The fermented food in the Philippines. (Part II). Basi (sugarcane wine). *J. Agric. Soc.* (Japan) 22 : 118 – 133.

Tanuwidjaja, L. (1972). Sticky rice fermentation (tape ketan fermentation). *Bull. L. K. N. K-12.* Indonesian Institute Sciences, National Institute Chem. Bandung.

Tegegne, Yetesha Work. (1957). *Ethiopian Ethological Society Bulletin.* No. 7. December.

Theivendirarajah, K., Dassanayake, M. D. and Jeyaseelan,K. (1977b). Studies on the fermentation of Kitul (Caryota Urens) sap. Symposium on Indigenous Fermented Foods, Bangkok, Thailand.

Toyama, N. (1976). Feasibility of sugar production from agriculture and urban cellusic wastes with *Trichoderma viride* cellulose. *Biotechnol. Bioeng. Symp.* 6: 207-219.

Tsujisaka, Y.; Lwai, M. and Tominaga, Y. (1972). A comparative study on some properties of fungal lipases. Proc. IV Int. *Ferment. Symp.* pp. 315 – 320.

Tsujisaka, Y.; Fukumoto, J. and Yamamoto, T(1958). *Nature*, 181, 770.

Unnikrishnan, K. R. and Bhattacharya, K. R. (1987). Influence of varietal difference on properties of parboiled rice. *Cereal Chem*: 64(4): 3115-321.

Uyenco, F. R and Gacutan, R. Q. (1977b). *Philippine tapuy.* Symposium on indigenous Fermented Foods, Bangkok, Thailand.

Van Uden, N. (1984). *Adv. Microb. Physiol.* 25 : 195 – 251.

Van Uden, N. and Leao, C. (1985). *Appl. Microbiol. Biotechnol.* 22 : 359 – 363.

Van Veen, A. G. (1972). Fermented rice foods. In: Houston, D. F. (ed). Rice: Chemistry and Technology. Am. Assoc. *Cereal Chem.*: St. Paul, MN.

Van Veen, A. G. , and Steinkraus, K. H. (1970). Nutritive value and wholesomeness of fermented foods. J. Agric *Food Chem.* 18:576.

Vander Watt, J. P. (1956). Kaffircorn malting and brewing studies. 11. Studies on the microbiology of Kaffir beer. *J. Sci. Food Agric.* 7:105-113.

Variyar , P. S. and Bandyopadhyay , C. (1995). Synthesis of 3, 4- dihydroxy – 6 – (N- ethyl- amino) benzamide, a new phenolic compound Isolated from green piper nigrum I) berries. *Indian Journal of Chemistry* 34 B; pp. 911 – 913.

Variyar, P. S. and Bandyopadhyay, C. (1994). Estimation of phenolic Compounds in green Pepper Berries (Piper Nigrum L.) by high performance liquid Chromatography. *Chromatographia* 39: (11/12)

Variyar, P. S. , Pendharkar, M. B, Bannerjee, A. and Bandyopadhyay, C. (1988). *Phytochemistry*, 27. 715.

Variyar, P. S. ; Bandyopadhya, C. and Thomas, P. (1998). Effect of Gamma- Irradiation on the phenolic acids of some Indian spices. *International Journal of Food Science and Technology* 33: 533-535.

Variyar, P. S. ; Pendharkar, M. B. ; Banerjec, A. and Bandyopadhyay, C. (1998). Blackening In Green Pepper Beries. *Phytochemistry*. 27(3): 715-717.

Virtanen, T. ; Autio, K. ; Suortti, T. and Poutanen, K. (1993). Heat- induced changes in native and acid modified oat starch pastes. *J. Cereal Sci.* 17; 137-145.

Vogel, S. and Gobezie, A. (1977). *Ethiopian tej.* Symposium of Indigenous Fermented Foods, Bangkok, Thailand.

Von Holdt, M. M. and Brand, J. C. (1960a). Kaffircorn malting and brewing studies VI. Starch content of Kaffir bear brewing materials. *J. Sci. Food Agric.* 11: 462-467.

Von Holdt, M. M. and J. C. Brand. (1960b). Kaffircorn Malting and brewing studies VII. Change in the carbohydrates of Kaffircorn during malting. *J. Sci. Food Agric.* 11: 467-471.

Wai, N. (1968). Final Technical Report Public Law 480 project UR – A6 –(40) –1, 89 pp.

Wai, N. (1929). *Science* 70, 307.

Wang, H. L. and Hasseltine, C. W. (1970). Sufu and Lao-Chao. *J. Agric. Food Chem.* 18 :572 – 575 .

Wang, H. L.; Vespa, J. B. and Hesseltine, C. W. (1974). Acid protease production by Fungi used in soybean Food Fermentation. *Appl. Microbiology* 27 (5) : 906 – 911.

Wang, H. L. and Hesseltine, C. W. (1965). Studies on the extracellular proteolytic enzymes of *Rhizopus oligosporus. Can . J. Microbiol.* 11 : 727 – 731.

Wang, H. L. ; Swain, E. W. and Hesseltine, C. W. (1975). Mass Production of *Rhizopus oligosporus* Spores and their application in Tempeh Fermentation. *Journal of Food Science* 40 : 168 – 170.

Wang, H. H. (1978a). Two types of models on solid state fermentation involved in mushroom cultivation for prediction of biomass formation. *Mushroom Sci.* 10: 1039-1041.

Wang, H. H. (1980). Fermented rice food. In B. S. Luh, (ed.)*Rice: production and utilization*, Westport, CT: AVI, pp. 650-689.

Wang, H. H. (1978b). The effect of CO_2 on nuclear selection of fungal mycelium in the solid-state fermentation. Abstracts of contributed paper, 3rd International symposium on the Genetics of Industrial Microorganisms. June 4-9, 1978, at the Univ. of Wisconsin, Madison, WI. p. 14.

Wang, H. H. (1989). Utilization of particulate agricultural products through solid state fermentation. *Proceedings of the National Science Council (ROC)* part B 13(3):145-159.

Wang, H. H. ; Chiou, C. M. and Chang, Y. M. (1970). Some propertiesof mutants *Rhizopus spp.* Induced by UV Irradiation. *Chin. J. Microbiol.* 3: 105-110.

Wang, H. H. ; Chiou, C. M. ; Ma, D. P. ; Lin, M. C. and Huang, S. Y. (1974). CO_2 level inside the composts and fructification of*Agaricus biosporus. Mushroom Sci.* 9:20-22.

Wang, H. H. ; Shih, T. H. ; Chion, C. M. ; Fan, T. I. and Chuang, H. L. (1968). A short method for the determination of aflatoxin in wine products. *J. Chin. Agric. Chem. Soc.* 6(3-4): 69-75.

Wang, H. L. (1967). Release of proteinase from mycelium of *Mucor hiemalis. J. Bacteriol.* 93: 1794 – 1799.

Wang, H. L. (1967a). *Food Technol.* 21, 15.

Wang, H. L. ; Ruttle, D. I. and Hesseltine, C. W. (1967). *Soc. Exptl. Biol. Med.* 31, 579 (1967).

Wang, H. L. ; Ruttle, D. L. and Hesseltine, C. W. (1969). Milk-clotting of proteinasesproduced by *Rhizopus. Can. J. Microbiol.* 15: 99-104.

Wang, H. L. ; Vespa, J. B. and Hesseltine, C. W. (1974). Acid protease production by fungi used in soyabean food fermentation. *Appl Microbiol* 27(5):906-911.

Wang, H. L. (1967). Release of protienase from mycelium of *Mucor hiemalis*. *Journal of Bacteriology*. 1794-1799.

Wang, L.; Wang Y. J. and Porter, R. (2002). Structures and physiochemical properties of wild rice starches. *J. Agric. Food Chem.* 50:2695-2699.

Wang,H. L. (1967b). *J. Bacteriol.* 93, 1794.

Weimberg, R. and Orton, W. L. (1966). Elution of exocellular enzymes from *Saccharomyces fragillis* and *Saccharomyces cerevisiae*. *J. Bacteriol.* 91: 1-13.

Wheals, A. E. ; Basso, L. C. ; Alves, D. M. and Amorim, H. V. (1999). Fuel ethanol after 25 years. *Trends Biotechnol* 17: 428-487.

Williams, J. and Bevenue, A. (1953). A note on sugar in rice. *Cereal Chem.* 30: 267-269.

Wong, P. W. and Jackson, H. (1977). Fermented foods of Sabah (Malaysia). Symposium on Indigenous Fermented Foods. Bangkok, Thailand.

Wood, B. J. B. (1981). Introduction of new fermented foods into Western culture. In: *Advances in Biotechnology*, edited by M. Moo-Young. Willowdale, Ontario, Canada: Pergamon Press Canada, pp. 467-472.

Wulf, L. W. and Nagel, C. W. (1976). Analysis of phenolic acid and flavonoids by HPLC. *J. Chromatogr.*, 116, 271-27

Yabuki, M. ; Ono, N. ; Hoshino, K. and Fukui, S. (1977). Rapid induction of á-amylase by non growing mycelia of *Aspergill oryzae*. *Appl Environ Microbiol* 34: 1-6.

Yashizawa, K. (1977). Traditional fermentation Industries in Japan. Paper read at a symposium of indigenous fermented foods in conjunction with the 5th Conference of Global impacts of Applied Microbiology, Bangkok, Thailand, Nov. 21-27.

Yemn, E. W. and Cooking, E. C. (1955). The determination of amino acids with ninhydrin, *Analyst.* 80: 209-213.

Yen, G. C. and Chen, H. Y. (1995). Antioxidant activity of various tea extracts in relation to their antimutagenicity. *J. Agric. Food Chem.* 43, 27-32.

Yeoh,Quee Lan. (1977b). Malaysian tapai. Symposium on Indigenous Fermented Foods, Bangkok, Thailand.

Zhou, Z. ; Robards, K. ; Helliwell, S. and Blanchard, C. (2002). Composition and functional properties of rice. *International Journal of Food Science and Technology* 37: 849-868.

Zoecklein, B. W. ; Fugelsang, K. C. ; Glump, B. H. and Nury, F. S. (1995). *Wine Analysis and Production:* chapman and Hall: New York.

Index

■■■